COGNITIVE GADGETS

COGNITIVE GADGETS

The Cultural Evolution of Thinking

Cecilia Heyes

THE BELKNAP PRESS OF
HARVARD UNIVERSITY PRESS
Cambridge, Massachusetts • London, England
2018

Printed in the United States of America

First printing

Library of Congress Cataloging-in-Publication Data

Names: Heyes, Cecilia M., author.
Title: Cognitive gadgets : the cultural evolution of thinking / Cecilia Heyes.
Description: Cambridge, Massachusetts : The Belknap Press of Harvard University Press, 2018. | Includes bibliographical references and index.
Identifiers: LCCN 2017041745 | ISBN 9780674980150 (hardcover : alk. paper)
Subjects: LCSH: Cognition and culture. | Nature and nurture. | Social evolution. | Evolutionary psychology.
Classification: LCC BF311 .H46916 2018 | DDC 155.7—dc23
LC record available at https://lccn.loc.gov/2017041745

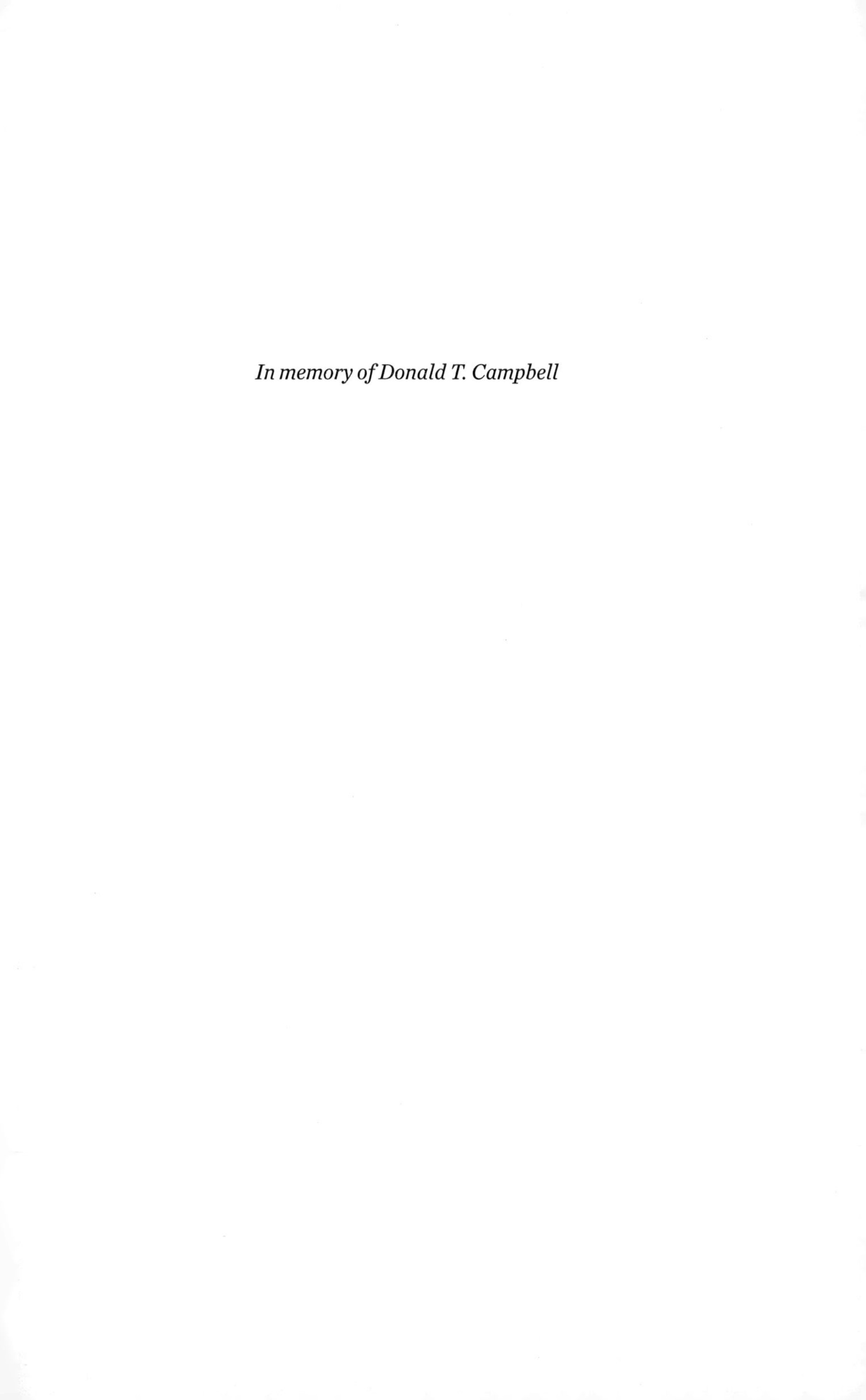

In memory of Donald T. Campbell

Contents

COGNITIVE GADGETS

Introduction

WHAT MAKES US SUCH PECULIAR ANIMALS? Compared with other creatures, we humans lead very strange lives. No other animals have so completely transformed their environment, become so dependent on cooperation for survival, and constructed, along the way, the vast edifices of knowledge and skill in which all human lives are embedded: technology, agriculture, science, religion, law, politics, trade, history, art, literature, music, and sports. Why? What is it about the human mind that enables us to live such unusual lives, and why do our minds work that way?

In this book I argue that the answer to these questions is "cognitive gadgets." We humans have created not just physical machines—such as pulleys, traps, carts, and internal combustion engines—but also mental machines; mechanisms of thought, embodied in our nervous systems, that enable our minds to go further, faster, and in different directions than the minds of any other animals. These distinctively human cognitive mechanisms include causal understanding, episodic memory, imitation, mindreading, normative thinking, and many more. They are "gadgets," rather than "instincts" (Pinker, 1994), because, like many physical devices, they are products of cultural rather than genetic evolution.[1] New cognitive mechanisms—different ways of thinking—have emerged, not by genetic mutation, but by innovations in cognitive development. These novelties have been passed on to subsequent generations, not via genes, but through social learning; people with a new cognitive mechanism passed it on to others through social interaction. And some of the new ways of thinking have spread through human populations, while others have died out, because the holders had more "students," not just more "babies" (Sober, 1991).

Psychologists often use gadgets as metaphors. They suggest that various aspects of the human mind operate in the same way as circuit boards, cisterns, search lights, search engines, thermostats, resistors, and the bristles of a Swiss Army knife. But, if I am right, the resemblance runs much deeper. Distinctively human ways of thinking are products of the same process—cultural evolution—as machines in the outside world; they are pieces of technology embodied in the brain. Genetic evolution has given humans more powerful general purpose mechanisms of learning and memory, tweaked our temperaments, and biased our attention so that it is focused on other people from birth. But—drawing on comparative and developmental psychology, cognitive neuroscience, philosophy, anthropology, behavioral economics, and theoretical biology—I argue in this book that it is the information we get from others, handled by general purpose mechanisms, that builds distinctively human ways of thinking.

The first three chapters lay some foundations for cultural evolutionary psychology. Chapter 1 says more about the cognitive gadgets theory—what it is, and what it is not—explaining how and why cultural evolutionary psychology builds on evolutionary psychology and cultural evolutionary theory. Chapter 2 draws on the philosophy of biology, arguing that, although we now know that some versions of the nature-nurture debate were deeply misguided, it is important to discover, for any particular feature of human cognition, the ways and extent to which the feature is shaped by: (1) genetically inherited information; (2) culturally inherited information; and (3) information derived directly from the environment in the course of development. Chapter 2 also includes an overview of contemporary cultural evolutionary theory, showing how it can be applied, not only to cognitive products ("grist"), but also to cognitive mechanisms ("mills"). Chapter 3 focuses on features of distinctively human cognition that have been shaped primarily by genetically inherited information. It surveys behavioral and neurological evidence that, far from being "blank slates," or just like the minds of chimpanzees, the minds of newborn human babies are equipped with high capacity mechanisms of learning and memory, species-specific attentional mechanisms, and a tendency to find social cues especially rewarding.

Chapter 4 examines the nature of cultural learning that enables cultural inheritance—the cultural analogue of DNA replication—and provides an introduction to the heart of the book, Chapters 5–8. Each of these chapters examines a type of cultural learning (selective social learning, imitation, mindreading, and language) and argues, from the available evidence, that its distinctively human characteristics depend on culturally inherited information. I focus on the mechanisms of cultural learning—the cognitive gadgets that enable humans to learn from others with extraordinary efficiency, fidelity, and precision—for two reasons. First, these distinctively human cognitive mechanisms are especially important because they are gifts that go on giving: culturally inherited skills that enable the cultural inheritance of more skills. Second, evolutionary psychologists and cultural evolutionists disagree about the origins of many cognitive characteristics, but both parties are convinced that the mechanisms of cultural learning are cognitive instincts, not cognitive gadgets. This consensus suggests that the mechanisms of cultural learning are the hardest nuts to crack—the cognitive mechanisms that are least likely to be explicable as products of cultural evolution.

Social learning is said to be "selective," or to involve "social learning strategies," when the impact on behavior of observing another agent varies with the circumstances in which the encounter occurs, or with the characteristics of the observed agent, or "model"—for example, when older models have more impact than younger models. In Chapter 5, I argue that most selective social learning—found in nonhuman animals, children, and adults—is due to domain-general learning and attentional processes, that is, to processes that have not been specialized for social interaction, let alone for cultural inheritance. However, a small proportion of social learning strategies, found only in adult humans, depend on explicit metacognition—on thinking about thinking. These, and only these, behavioral effects are genuinely "strategic," and genuinely examples of cultural learning. The evidence suggests that, like other explicitly metacognitive rules, these metacognitive social learning strategies are learned through social interaction—culturally, rather than genetically, inherited.

Imitation occurs when an observer copies the topography of a model's action; observing the way that parts of a model's body move *relative to one another* causes the observer to produce movements in which the parts of his or her own body move in a similar way. In Chapter 6, I agree with the century-old view that imitation is "special"—much more highly developed in humans than in any other species, and dependent on mechanisms that are not involved in other kinds of learning. I also agree that these mechanisms contribute to the fidelity of cultural inheritance. My rebellious streak comes out only in relation to the question of where imitation comes from. Offering an original theory of the mechanisms mediating imitation, and a wide range of empirical evidence in support of that theory, I argue that the capacity to imitate is acquired through sociocultural experience.

Mindreading involves the ascription of mental states, such as beliefs and desires, thoughts and feelings, to oneself and to others. In Chapter 7, I suggest that genuine mindreading contributes to cultural inheritance primarily by enhancing the effectiveness of teaching, but that many of the behavioral effects attributed to mindreading—the "implicit" or "automatic" effects reported in apes, infants, and adults under time pressure—are not genuine cases of mindreading; they are due to domain-general psychological processes. These processes can generate predictions about behavior that simulate the effects of mindreading, and when they do, the agent may be described as "submentalizing." Where does real mindreading come from? From the same kinds of conversation-based social interactions that support the development of print reading or literacy. It is culturally inherited.

No one doubts that language—communication using words or signs in a structured and conventional way—is a hugely important form of cultural learning. When it comes to language, the crucial question is not whether it is a form of cultural learning, but where the language faculty originated: genetic or cultural evolution. In Chapter 8, I approach this debate as an outsider—neither a linguist nor a language scientist—and with an open mind. Indeed, it would have been convenient for the purposes of this book if I had found in the language debate a compelling case for an innate language faculty; a rock-solid cognitive instinct on which cultural evolution had constructed cognitive gadgets. But that is not what I found. Insofar as the two ideas can be tested against one another, I find the case for the cultural evolution of language at least as strong as the genetic alternative.

The core chapters, Chapters 5–8, have particular selling points. Chapter 5 addresses very directly a question which cultural evolutionists have tended to avoid: Exactly what is it about selective social learning that promotes cultural evolution? Chapter 6, on imitation, looks in detail at how a new cognitive mechanism can be constructed by domain-general cognitive processes through social interaction. Chapter 7 presents an innovative view of mindreading, offering an alternative to the long established nativist and theory-theory perspectives. Chapter 8 gives an informed but dispassionate overview of the current status of the debate about the origins of language; I have read widely, but I don't have a dog in that fight.

All of the case studies are unusual in bringing to the cultural-evolutionary table theory and evidence, not only from primatology and developmental psychology, but from experimental psychology and cognitive neuroscience.

The final chapter takes a step back to consider how the cognitive gadgets theory measures up against the chronology of human evolution, and what it implies about human nature. Cultural evolutionary psychology implies that human minds are more agile, but also more fragile, than was previously thought. We are not stuck in the Pleistocene past with Stone Age minds, and well-targeted educational interventions have the potential to transform cognitive development, but we have more to lose. Wars and epidemics can wipe out not just know-how, but the means to acquire that know-how. The cultural evolutionary perspective also has disciplinary implications. It does not suggest, as have many evolutionary psychologists, that all research on human minds and human lives must be informed by evolutionary theory. On the contrary, it suggests that research on the developmental and evolutionary origins of human cognition should be informed by the humanities and social sciences.

1

A QUESTION AND MANY ANSWERS

IN THIS CHAPTER, I FIRST OUTLINE HOW THE COGNITIVE gadgets theory differs from other, recent answers to the question "What makes us peculiar?" I locate my answer within the "logical geography" (Ryle, 1945) of contemporary research on human evolution. Then I explain how and why the cognitive gadgets theory builds on some of the other answers, and suggest that the origins of literacy provide a proof of principle for cognitive gadgets.

LOGICAL GEOGRAPHY

Dimensions

Some political and religious groups hold that the differences between humans and other animals are due to supernatural forces and provide moral justification for privileging human interests over all others. In contrast, contemporary scientific enquiry assumes that any differences between humans and other animals, in degree or in kind, result exclusively from natural processes—many of them evolutionary—and while the differences may inform debate about the ethical treatment of animals, their moral implications are far from self-evident. When scientists ask "What makes us peculiar?" they are not assuming that humans are the only peculiar animals, or that our peculiarities make us morally superior.

By definition, all scientific inquiry about human distinctiveness aims ultimately to explain the manifest differences between our lives and the lives of other animals—the differences in geographical distribution, habitats, and habits that the proverbial Martian would be able to detect while visiting Earth. Scientists look for explanations in a variety of places. Some research focuses on our bodies, for example, tracing the effects of bipedalism and our remarkable manual dexterity. Other work zeros in on the brain, emphasizing that our brains are unusually large, relative to our bodies, and that certain parts have expanded more than others in the course of human evolution. A third focus is on behavior, the things people do. Perhaps a key to understanding human distinctiveness lies in our social behavior, use of tools, or control of fire. The final major focus of evolutionary, scientific inquiry about human distinctiveness is on the mind. Research of this kind tries to identify the mental processes, or ways of thinking, that make humans special.

Of course, these four foci—bodies, brains, behavior, and the mind—are complementary, not competitive. The mind is implemented in the brain, the brain is part of the body, the mind-brain controls behavior, and behavior is enacted by the body. Ultimately, therefore, a full explanation of the peculiarities of human lives must integrate research with all four foci.

Evolutionary answers to the question "What makes us peculiar?" also vary in the extent of their preoccupation with history and forces. Answers that are high on the historical dimension, "narrative theories," offer a sequence and chronology of key events in human evolution. For example, they link major changes in brain structure and behavior with climactic or demographic events that may have provoked those changes. Answers that are high on the forces dimension, "force theories," are concerned with the processes involved in human evolution: cultural inheritance, epigenetic inheritance, gene-culture coevolution, genetic assimilation, genetic drift, and niche construction, as well as natural selection operating on genetic variants. (All of these processes will be discussed in later pages.) The ideal theory would be synthetically high on both historical and force dimensions—it would use chronology as evidence of forces, and forces to explain chronology. There are already some admirably synthetic theories (for example, Sterelny, 2003; 2012), but synthesis is a very tall order. The patchiness of the fossil and archaeological records compels narrative theories to be speculative (they are often dismissed as "just-so stories"), and when detailed narrative is combined with analysis of forces, the resulting story can be intractably complex and untestable.

Evolutionary Psychology

Where does the cognitive gadgets answer lie in this map, defined by foci and preoccupations? How does it compare with the neighbors? The focus of the cognitive gadgets answer is firmly on the mind. In agreement with what is known as "evolutionary psychology"—or, to distinguish it from other evolutionary approaches to the study of mind and behavior, as "High Church evolutionary psychology" (Barkow, Cosmides, and Tooby, 1995; Pinker, 1994)—I am convinced that relationships between the brain, behavior and the world cannot be understood satisfactorily without a middleman—that is to say, without describing those relationships at an abstract, mental level.

Imagine someone hunting in the wilderness with a spear. It may be possible, with the help of immensely complex mathematical models, to document what typically happens in the hunter's brain whenever there is a change in the pattern of light entering his eyes as he scans the horizon. It may even be possible to correlate these light-related changes in neural firing with the hunter's behavior: Type A patterns predict he will continue scanning; Type B patterns predict he will crouch closer to the ground; and Type C patterns indicate he will raise his spear. But this huge, complicated matrix of inputs, brain activities, and outputs would not make sense of the hunter's action, or provide information about what he is likely to do under different circumstances, unless it were translated into mental terms; for example, into a description of what the hunter "sees," "misses," "wants," and "knows." Terms like "seeing" and "knowing" refer to the kinds of activities and relations documented in the hunter's correlation matrix. They do not refer to "extra things" done by the brain or by a spooky mental substance. However, without abstraction of the kind that mental terms provide, behavioral science and neuroscience provide information without insight, and precision without predictive power.

In common with many evolutionary psychologists, I believe some of the most effective abstractions come not from folk psychology ("seeing" and "knowing") but from cognitive science. "Folk psychology" refers to the blend of wisdom and old wives' tales that we use to talk about the mind in everyday life. It explains behavior with reference to the thoughts and feelings, beliefs and desires, of whole agents. For example: "Nebeela nodded her head because she *wanted* to bid for the Miro, and *believed* the auctioneer would *understand* her head movement to be a bid." The term "cognitive science" has been used since the early 1970s to refer to interdisciplinary scientific research on the mind. The disciplinary mix includes experimental psychology, computer science, linguistics, neuroscience, and the philosophy of mind. Many of the explanations offered by cognitive science liken the mind to a computer, cast thinking as "information processing," and are pitched at a "sub-personal" level (Dennett, 1969; 1987). That is, in contrast with folk psychology, which takes mental states of the whole person (for example, beliefs and desires) to be the drivers of behavior, cognitive science often explains behavior as deriving from the activities of parts of the mind, and of the interactions among these parts. For example: "Nebeela said 'blue' when she saw BLUE written in green ink because two parts of her mind—one responsible for naming colors, and the other for reading words—competed for control of Nebeela's speech mechanisms, and the reading part won the contest." The sub-personal explanations offered by cognitive science are not familiar or intuitive, but they burrow deeper into the mind than folk psychology, and many have survived rigorous experimental tests.

Introducing cognitive science to inquiry about human evolution, and vice versa, was one of the primary purposes of evolutionary psychology and is likely to be its most enduring achievement. The forerunner of evolutionary psychology, human sociobiology (Wilson, 1975), attempted to explain the evolution of distinctively human social behavior, especially altruistic behavior, either without mentioning minds at all, or while relying on casually generated, folk-psychological characterizations of how the mind works. Similarly, the still-thriving field of human behavioral ecology (Cronk, 1991; Laland and Brown, 2011) largely ignores the mind as it uses mathematical modeling to investigate whether distinctively human behaviors, especially foraging behaviors, are likely to be adaptive (to enhance reproductive fitness) and, if so, to be adaptations (to have evolved because they enhance reproductive fitness). Adopting a strategy that has been described as the "phenotypic gambit" (Grafen, 1984; Hoppitt and Laland, 2013) and "blackboxing" (Heyes, 2016a), human behavioral ecologists sometimes allude to the brain, but very rarely to the mind, and when they do say something about the mind, they usually reach for folk psychology. Consequently, although sociobiology and human behavioral ecology have provided valuable information about human evolution, I think evolutionary psychologists were right to point out, some twenty years ago, that it was high time the mind was taken more seriously.

The cognitive gadgets answer is at one with evolutionary psychology in focusing on the mind and emphasizing the importance of cognitive science in explaining human distinctiveness. However, the cognitive gadgets answer is completely at odds with evolutionary psychology—or at least with the part known as "High Church evolutionary psychology"—in its claims about the forces that have shaped, and continue to shape, the human mind. Most evolutionary psychologists argue, assert, or merely assume that genetic evolution is the architect of the human mind. According to this "cognitive instincts" view, distinctively human ways of thinking are "in our genes." A newborn human baby does not enter the world understanding causality, with a full supply of mental maps, and talking in complete sentences, but she contains in her genes very specific programs for the development of these capacities; programs that are capable of building distinctively human cognitive mechanisms—such as causal understanding, mental mapping, and language—with minimal help from learning. The environment in which a child grows up is seen as "triggering" or "evoking" cognitive development; not, as it is in the cognitive gadgets theory, as forging or constructing distinctively human ways of thinking.

Cultural Evolutionary Theory

Cognitive gadgets is a "force theory" rather than a "narrative theory." It is consistent with what is known about the chronology of human evolution (see Chapter 9), but it is primarily concerned with the processes that have shaped the human mind and regards learning—especially social learning—and cultural evolution as dominant among these processes. In emphasizing the importance of social learning and culture in explaining why humans live such peculiar lives, the cognitive gadgets theory is akin to "cultural evolutionary theory" (Lewens, 2015).

Contemporary cultural evolutionary theory emerged in the 1980s (Boyd and Richerson, 1988; Cavalli-Sforza and Feldman, 1981), building on ideas formulated in the preceding twenty years (Campbell, 1965; 1974), and resolutely rejecting the politically stained "social Darwinism" of the Victorian era. It is now a broad church (see Chapter 2). Cultural evolutionists are united in believing that evolution—defined as a change in the distribution of characteristics within a population over time—can be powered not only by genetic inheritance but by cultural inheritance. Characteristics, or "traits," can increase or decrease in frequency, not only as they become more or less likely to be passed on to biological descendants via genetic mechanisms, but also as they become more or less likely to be passed on to cultural descendants, who may or may not be genetically related to their cultural parents, through social interaction.

The crucial difference between contemporary cultural evolutionary theory and the cognitive gadgets theory concerns the traits they have in their sights. Until now, cultural evolutionary theory has been applied to observable behavior and artifacts. For example, it has been used to explain change over time in the frequency of people in a population who have a small family, or who use a particular kind of fish hook. In contrast, the cognitive gadgets theory applies cultural evolutionary theory to the mechanisms of thought—the mental processes that generate and control behavior. For example, it seeks to explain change over time in the frequency of people in a population who are capable of calculating a shortcut across unexplored territory (mental mapping), who can entertain a theory about how an instrument works (causal understanding), or who have cognitive equipment allowing them to copy facial expressions (imitation). The cognitive gadgets answer is concerned not with the grist of the mind—what we do and make—but with its mills, the way the mind works (Aquinas, 1272; Heyes, 2012a).

When questions arise about the origins of distinctively human cognitive mechanisms, even the most enthusiastic and pioneering cultural evolutionists begin to look remarkably like High Church evolutionary psychologists. They may deny with great resolution, and on the basis of sophisticated mathematical models, that genetic mechanisms are responsible for the evolution of human behavior (grist), while assuming without comment, or on the basis of meager evidence, that distinctively human cognitive mechanisms (mills) have been fashioned by the genes.

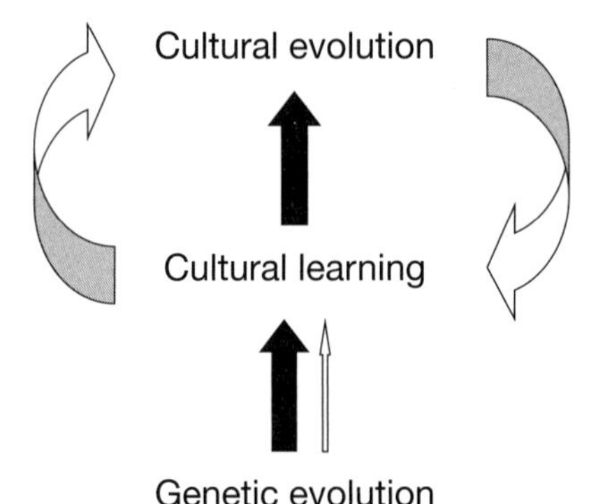

1.1 Relations between genetic evolution, cultural learning, and cultural evolution. Evolutionary psychology and cultural evolutionary theory imply that genetic evolution has produced cultural learning and that cultural learning enables cultural evolution (dark arrows). In contrast, cultural evolutionary psychology suggests that genetic evolution has made only a modest contribution to the emergence of cultural learning. The mechanisms that enable cultural evolution are themselves products of cultural evolution (light arrows).

This juxtaposition is particularly striking when it comes to the cognitive mechanisms that enable cultural inheritance—mechanisms known collectively as "social learning" or "cultural learning" (see Chapter 4). The picture painted by contemporary cultural evolutionary theory has the virtue of simplicity (Figure 1.1). Genetic evolution has given humans mechanisms for cultural learning, and, using these cognitive instincts, we learn from others most of what we need for survival and reproduction in the particular geographical area, and particular social group, in which we live. It is a simple picture, but I argue that it does not line up with the evidence from cognitive science. That evidence suggests that the distinctively human cognitive mechanisms involved in social learning are not only processes but also products of cultural evolution.

In summary (Figure 1.2): the cognitive gadgets answer to the question "What makes us peculiar?" is like evolutionary psychology in targeting the mind and drawing on cognitive science, and like cultural evolutionary theory in emphasizing the importance of social learning as a force in human evolution. However, it is quite different from both

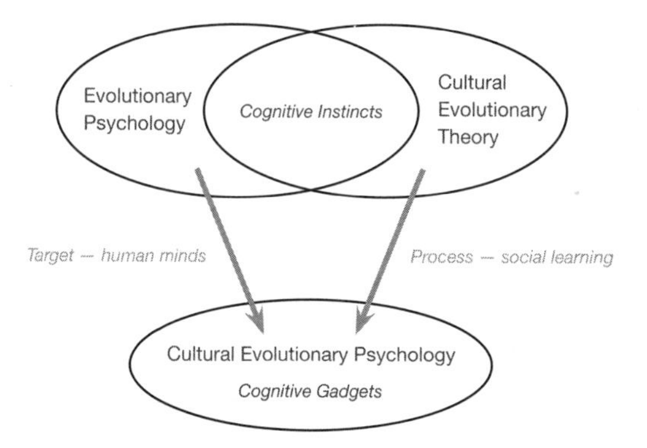

1.2 Relations between evolutionary psychology, cultural evolutionary theory, and cultural evolutionary psychology.

of these approaches in suggesting that distinctively human cognitive mechanisms are gadgets rather than instincts; products of cultural rather than genetic evolution. Given these similarities and differences, the cognitive gadgets theory represents a progression from evolutionary psychology and cultural evolutionary theory that I call "cultural evolutionary psychology."

Fellow Travelers

Although the cognitive gadgets theory is at odds with evolutionary psychology and cultural evolutionary theory on the origins of distinctively human cognitive mechanisms, I most certainly am not a lone voice in the wilderness crying "Culture changes the way we think." For many social anthropologists, especially those who identify themselves as "cultural psychologists" (Shweder and Sullivan, 1993), this is a fundamental tenet of their work. They regard all aspects of the mind as interdependent with the cultural environment in which it develops, to the point where mind-and-environment are inseparable. This conviction, among researchers who have a rich body of direct experience with people from non-Western cultures, provides encouragement for the cognitive gadgets hypothesis. However, cultural evolutionary psychology departs from cultural psychology, and from the approach adopted by most social anthropologists, in being rooted in cognitive science and evolutionary theory. For example, it takes a keen interest in characteristics that are typical of humans, as well as variation between individuals and groups. Also, cultural evolutionary psychology does not seek to abolish the distinction between nature and nurture (see Chapter 2), or to revise our everyday ontology in which minds continuously interact with, but are distinguishable from, the world around them (Greenwood, 2015).

Other voices saying "Culture changes the way we think" come from "cross-cultural psychology." In this field, where many researchers draw on cognitive science, experiments in which people from different cultures are given the same behavioral tasks reveal both species-typical human psychological characteristics and fascinating patterns of between-group variation (Haun, Rapold, Call, Janzen, and Levinson, 2006; Nisbett, 2010; Shiraev and Levy, 2014; Winawer et al., 2007). Cross-cultural psychology is, arguably, a methodology rather than a research program with its own theoretical framework. Consequently, there is significant potential for synergy between cultural evolutionary psychology and cross-cultural psychology, with the former providing theory and the latter, evidence.

Within the logical geography of research on human distinctiveness, the theories that lie closest to the cognitive gadgets theory are those of Barrett (2017), Dennett (1991), Karmiloff-Smith (1995; 2015), and Tomasello (2009; 2014). Barrett's "theory of constructed emotion" has a different focus—emotion rather than cognition—and does not invoke cultural evolution specifically, but, like cognitive gadgets, it is rooted in cognitive science and underlines the importance of social interaction in shaping human minds. Dennett's "multiple drafts" account of consciousness implies, but does not state explicitly, that cultural evolution can shape cognitive mechanisms as well as cognitive products. Karmiloff-Smith's theory of "representational re-description" makes no reference to cultural evolution but is a pioneering attempt to specify, within a cognitive science framework, how sociocultural experience could produce new cognitive mechanisms. Finally, like the cognitive gadgets theory, Tomasello's "shared intentionality hypothesis" is a direct answer to the question "What makes us so peculiar?" focusing on the psychological processes involved in cultural inheritance, and emphasizing the importance in cognitive development of learning through social interaction. However, the shared intentionality hypothesis seeks a single psychological source of human distinctiveness ("shared intentionality"), rather than a set of distinctively human cognitive processes (mental mapping, causal understanding, imitation, etc.); it is theoretically and empirically rooted in a Vygotskian psychology, rather than cognitive science; and it appears to assume that humans genetically inherit highly specific propensities for cultural learning. If this is correct, unlike cultural evolutionary psychology, the shared intentionality hypothesis implies that w*hat* we think depends on cultural learning, but the *way* we think depends on our genes.

WHY NOW?

At the heart of the cognitive gadgets theory, of cultural evolutionary psychology, is the idea that social interaction in infancy and childhood produces new cognitive mechanisms; it changes the way we think. Until recently, this idea would have been very difficult to evaluate, since there was minimal contact between research on social interaction and cognitive mechanisms. Social psychologists examined how groups and individuals interact but took little interest in cognitive mechanisms. On the other hand, cognitive psychologists were keenly interested in cognitive mechanisms but applied their theories to asocial problems—for example, how a person who is (or might as well be) alone in a room sees an object, remembers the past, types on a keyboard, or finds the door. Since the 1990s, this situation has changed rapidly and radically. The field of "social cognitive neuroscience"—which combines social psychology, cognitive psychology, and developmental psychology, with lashings of brain imaging—is now one of the most vibrant and richly funded research enterprises in the natural sciences (Blakemore, Winston, and Frith, 2004; Lieberman, 2007). Social cognitive neuroscience makes it possible, for the first time, to take a really close look at the mechanisms responsible for online control of social interaction and, crucially, at the role of social interaction in their development.

For example, social cognitive neuroscientists have shown, using a combination of mathematical modeling and brain imaging, that many of the mechanisms involved in human social learning (Behrens, Hunt, Woolrich, and Rushworth, 2008) and teaching (Apps, Lesage, and Ramnani, 2015) have deep evolutionary roots. At the core of human social learning and teaching are the same, basic mechanisms used by all vertebrates, and many invertebrates, to learn about predictive relationships between events; the same "associative" or "statistical" mechanisms that enable bees to learn that flowers of particular colors yield more nectar, and birds to learn that, if their beak strikes a mollusk at a certain angle, with a certain force, the shell will open. This implies that, in many cases, distinctively human ways of thinking are not as brand new and shiny as we thought; they are made of old parts and, therefore, tweaking by genetic evolution, rather than heavy lifting, would be enough to get their development going (see Chapter 3).

Reading, or literacy, was studied by cognitive psychologists long before the eruption of social cognitive neuroscience, but brain imaging has confirmed dramatically that training a person to read reconfigures his or her cognitive system. This transformation is a proof of principle for cultural evolutionary psychology: it demonstrates that new cognitive mechanisms, and, specifically, new mechanisms of cultural learning, can be produced by cultural evolution (Heyes, 2012a). First, reading is an immensely powerful and distinctively human form of cultural learning; a cognitive process that enables those who are literate to access a huge store of information acquired by previous generations. Second, it is clear that reading has been made possible by cultural evolution. Written language emerged only five to six thousand years ago, too recently in human history for the genetic evolution of cognitive mechanisms dedicated to reading.

Research on the cognitive mechanisms involved in reading is informed by experiments measuring the speed of processing and the kinds of errors made during reading by healthy literate people, and by people with various types of brain damage. According to one of the most prominent theories, the "dual route cascaded model," the data from these studies imply that each competent reader has distinct cognitive routes from seeing a letter string to reading it aloud (Figure 1.3; Coltheart, Rastle, Perry, Langdon, and Ziegler, 2001). Each of these routes (indicated by arrows), and some of their component processes (indicated by boxes), is constructed by the process of learning to read. Even when a component is in place prior to literacy training, route construction transforms the way it operates.

Illustrating the scale of the changes induced by literacy training, brain imaging has shown that viewing written sentences activates large areas of the cortex more strongly in literate than in illiterate adults (Dehaene et al., 2010). These areas include: the right occipital cortex, which is involved in relatively low-level visual processing, and a focal area of the occipitotemporal cortex. The latter area is known as the visual word form area (VWFA) because it responds so reliably, in literate people, to the presentation of written words. If one did not know that reading is culturally inherited, it would be easy to mistake the reliable responding and precise localization of the VWFA for signs that the capacity to read depends on a cognitive instinct or an "innate module."

So, learning to read has major, constructive effects on the neurocognitive system. It does not, of course, create a new system from scratch. Like other genetic and cultural processes of adaptation, learning to read takes old parts and remodels them into a new system. The old parts are computational processes and cortical regions

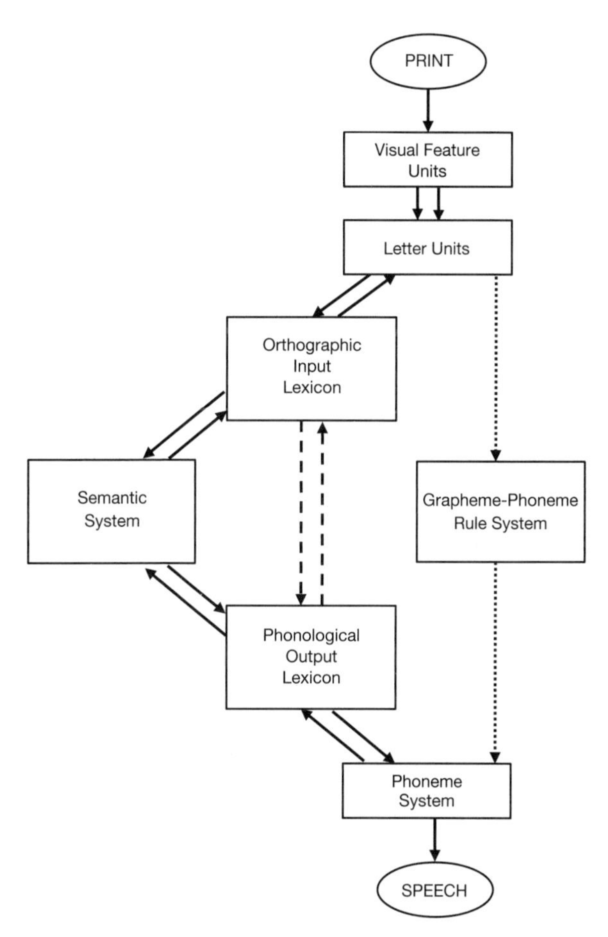

1.3 The dual-route cascaded model of reading aloud. The "lexical semantic route" (solid arrows) first goes from letters to a mental dictionary of printed word forms (orthographic input lexicon), then to a semantic system encoding word meanings, then to a phonological output lexicon, storing sound information relating to words, and finally to the system producing spoken words. The "lexical non-semantic route" (dashed arrows) bypasses the semantic system but uses the orthographic input and phonological output lexicons. The "grapheme–phoneme route" (dotted arrows) bypasses even these, allowing visually presented letters to activate phonemes and to produce speech directly. (Adapted with the permission of the American Psychological Association from Coltheart, Rastle, Perry, Langdon, and Ziegler, 2001.)

originally adapted, genetically and culturally, for object recognition and spoken language, but it is a developmental, cultural process—literacy training—that makes them into a new system specialized for cultural learning.

The case of reading shows clearly that distinctively human cognitive mechanisms, including mechanisms of cultural learning, can be culturally inherited. Cultural evolutionary psychology asks: How far does this go? To what extent are other distinctively human cognitive processes constructed and inherited through social interaction?

CONCLUSION

The cognitive gadgets theory, or cultural evolutionary psychology, addresses the question: What makes human lives so peculiar? It is a force theory rather than a narrative theory of human evolution, akin to evolutionary psychology in focusing on the mind, and to cultural evolutionary theory in emphasizing the importance of social learning and culture in shaping human characteristics. However, cultural evolutionary psychology makes a radical departure from both evolutionary psychology and cultural evolutionary theory in proposing that distinctively human cognitive mechanisms—ways of thinking—have been built by cultural evolution. They are cognitive gadgets rather than cognitive instincts; pieces of mental technology that are not merely tuned but assembled in the course of childhood through social interaction. Some of the components and engines of construction are genetically inherited, but the designer of the human mind is natural selection acting on cultural, rather than genetic, variants. We are taught the thinking skills that make us peculiar. Those skills are not "in our genes." The recent emergence of social cognitive neuroscience makes cultural evolutionary psychology a timely development in research on human evolution, and research on the acquisition of literacy provides a proof of principle.

To take culture seriously, we need to rethink the distinction between nature and nurture. Laying some groundwork for cultural evolutionary psychology, Chapter 2 tackles the nature-nurture issue and takes a closer look at what is meant by cultural evolution.

2

NATURE, NURTURE, CULTURE

THE DEVELOPMENT OF EVERY ASPECT OF HUMAN behavior and cognition depends on a rich, turbulent stew of factors. Even the development of something as simple as fight-or-flight reactions to threatening stimuli depends on complex, looping interactions between a multitude of DNA sequences (genes), epigenetic markers, environmental inputs (such as threat stimuli, parental care, nutrition), and the current state of the neuroendocrine system that these resources have conspired to produce (Loman and Gunnar, 2010). The rich interactive complexity of developmental processes makes it absolutely clear that, in cognition as in other biological systems, there are no pure cases of nature or of nurture; no biological characteristic is caused only by "the genes" or only by "the environment."

The absence of pure cases has led some people to conclude that the nature-nurture debate should be abandoned, but this simply does not follow. The composition of a cake never depends solely on the ingredients, nor solely on oven settings, but it still makes sense to ask to what extent, and in what ways, the various ingredients and oven settings contribute to the cake's appearance, flavor, and texture. Similarly, although every biological characteristic is the product of nature *and* nurture, it remains coherent and important to ask, for any particular characteristic, to what extent and in what ways nature and nurture contribute to its development. Indeed, answering these questions is, arguably, one of the core tasks of developmental biology and developmental psychology.

Nonetheless, those who are skeptical about contrasting nature with nurture, and innate with acquired characteristics, have emphasized two genuine and complementary dangers. On the one hand, efforts to explain development run the risk of being overwhelmed by the sheer multifactorial complexity of developmental systems. Unless we have good reasons to distinguish some contributors as more important than others, explanations of development, like Lewis Carroll's fictional map with a scale of one mile to one mile, will be too unwieldy to provide insight or a basis for intervention. On the other hand, in our efforts to simplify, producing a map that selectively represents the important features of the developmental terrain, we must be careful not to privilege some factors over others arbitrarily. This happened often in the behavioral sciences of the twentieth century. As the pendulum swung from instinct theory (Kuo, 1922) to behaviorism (Watson, 1930), and back again to High Church evolutionary psychology, via classical ethology (Lorenz, 1965; Tinbergen, 1963) and sociobiology (Wilson, 1975), behavioral scientists first privileged inherited factors, then focused on environmental influences, and finally put the genes back in the ascendant. These historical changes were not whimsical. For example, the rise of behaviorism was fostered by political pressure for educational reform in the United States, and its decline was accelerated by Chomsky's critique of Skinner's theory of language. But it would be hard to argue that these historical shifts were wholly rational responses to new empirical discoveries, or to better ways of thinking about the interplay between inherited and environmental factors in cognitive development. They were more like Gestalt switches, in which scientists looking at cognitive development sometimes saw a duck of inheritance, and sometimes a rabbit of environmental control (Figure 2.1; Kuhn, 1962).

Building on previous work in biology (Lorenz, 1965; Maynard-Smith, 2000; Plotkin and Odling-Smee, 1981), and with a close eye on the "neo-nativism" typical of contemporary evolutionary psychology (Griffiths, 2017), some philosophers of science have found a way to avoid the rock of excess complexity and the hard place of arbitrary privilege. Their method uses a "teleosemantic" conception of information to distinguish more and less important contributors to development, and to compare the roles of inherited and environmental factors.[1] This chapter outlines the teleosemantic method and provides an overview of research on cultural evolution. It then explains how the most thoroughly evolutionary approach to cultural evolution—the selectionist approach—can be applied not only to behavior and artifacts (grist), but also to cognitive mechanisms (mills), and therefore how it can be used in cultural evolutionary psychology.

2.1 The mallard-rabbit illusion.

BIOLOGICAL INFORMATION

The concept of information is used promiscuously in contemporary biology and psychology. It is applied not only to whole organisms when they are perceiving, thinking, signaling, and using language, but to the firing patterns of single neurons, causal processes within cells, and, ubiquitously, to DNA sequences; genes are understood to "carry information," or to be a "code." In many such cases, to say "X carries information about Y" is to say no more than "X correlates with Y." In this "Shannon" sense of information (Shannon, 1949), smoke carries information about fire because the two tend to go together, and a neuron carries information about Jennifer Aniston if it typically fires during exposure to pictures of Jennifer Aniston but not to pictures of other celebrities (Quiroga, Reddy, Kreiman, Koch, and Fried, 2005). The Shannon conception of information can be very useful but, by itself, is not an adequate guide for research on the development of biological systems. In isolation, the Shannon conception would leave accounts of development cluttered with too many factors. The number and size of the rings in a tree trunk correlate with the conditions in which the tree developed, and therefore carry information, in the Shannon sense, about the tree's history, but the rings do not play a significant role in the tree's development. Similarly, as children develop precision grip—the capacity to hold and lift objects securely between thumb and forefinger—they go through a regular series of stages characterized by the relative timing of pinch and lift components of the movement (Forssberg, Eliasson, Kinoshita, Johansson, and Westling, 1991). These stages correlate with, and therefore carry Shannon information about, the age of the child, but it is an open question whether the relative timing of pinch and lift at one stage in development contributes to their relative timing, or to any other aspect of precision grip, at a later stage.

The teleosemantic conception of information builds on the Shannon conception (Millikan, 1984; Shea, 2013; Sterelny, Smith, and Dickison, 1996). At its heart, the teleosemantic view suggests that, of all the factors that correlate with developmental outcomes—and therefore carry Shannon information—the ones that *really* carry information (or "represent," "code" or "carry meaning") are those for which the correlation exists by virtue of a Darwinian selection process. More specifically: *a biological structure X carries information about Y only if the state of X correlates with Y, and X was selected (preserved and propagated to future generations) because its states correlate with Y.* On this teleosemantic view, Jennifer Aniston neurons do not carry information about Jennifer Aniston because Jennifer Aniston was not part of the environment in which these neurons evolved. She was not around millions of years ago when neurons, and basic mechanisms of neural plasticity, were evolving. Nor was she around hundreds of thousands of years ago when human-specific mechanisms of neural plasticity were evolving. Therefore, although the processes that contribute to the development of Jennifer Aniston neurons undoubtedly evolved, they did not evolve *because* they tend to produce Jennifer Aniston neurons. It is not their function or purpose to fire when images of Jennifer Aniston are present, and, consequently, Jennifer Aniston neurons do not, in the teleosemantic sense, represent or carry information about Jennifer Aniston.

The teleosemantic view helps explanations of cognitive development avoid the rock of excess complexity without hitting the hard place of arbitrary privilege. It points out that some contributors to development are more important than others from an evolutionary perspective. From a proximate causal perspective, exposure to images of Jennifer Aniston is very important in the development of Jennifer Aniston neurons; but from an evolutionary perspective—when our purpose is to explain the deep, historical roots of human cognition—Jennifer Aniston can be left out of the picture.[2]

The teleosemantic view has another major advantage: it gives us a common currency in which to compare inherited and environmental contributions to development (Shea, 2013). We are accustomed to thinking about environmental contributions in terms of information. When a water flea grows defensive armor after exposure to a chemical correlate of predators, it is natural to say that the flea "detected" or "found out" that predators were nearby. Similarly, when a cat starts to appear beside his bowl at the same time each day, it is natural to say that the cat "knows" it is feeding time. The teleosemantic view allows inherited contributions to development to be understood in the same, informational way as environmental contributions. For example, ostriches, which are born with calluses in places where their skin will touch the ground, can be said to "have information," or to "know," where to grow calluses. This is warranted by the teleosemantic view: ostriches have an inherited biological structure, in the genome, which has been favored by selection, because it promotes the embryonic development of calluses. In principle, the signal carrying information about where to grow calluses could come from the environment, and it probably did at some point in ostrich evolution, but this information now comes from the genome. It is an inherited, rather than an environmental, contribution to development.[3]

But the genome is not the only repository of inherited information, and the mechanisms that copy DNA sequences in biological reproduction are not the only means of passing information from one generation to the next (Jablonka and Lamb, 2005; Uller, 2008). In the last twenty-five years, it has become clear that, in a wide range of plant and animal species, information with stable effects on development can be passed from one organism to another epigenetically, that is, via changes in a chromosome that do not involve alterations in the DNA sequence. Yet more important for human evolution, cultural inheritance, or cultural evolution, has emerged as a major force in shaping what people know and how they behave (Henrich, 2015; Lewens, 2015;

Morin, 2015; Tomasello, 2009). Information that we inherit from others through social interaction (via certain kinds of social learning) has such powerful effects that this information—"culture"—ranks alongside "nature" (genetically inherited information) and "nurture" (information derived from direct interaction between the developing system and its environment) as a determinant of human development.

CULTURAL EVOLUTION

Cultural evolution comes in various strengths: historical, populational, and selectionist (Brusse, 2017; Godfrey-Smith, 2009). At its weakest, the term "cultural evolution" refers to any change over time in the "cultural" characteristics of a population; characteristics that make the population distinct from others of a similar kind. This very loose conception of cultural evolution is sometimes invoked by business managers and politicians but, by itself, is not used by contemporary scientists. It allows virtually any characteristic that varies between populations to be regarded as cultural, and it is called "historical" because it takes evolution to mean nothing more than "change over time."

The populational conception of cultural evolution is more interesting. It assumes that large-scale changes in, for example, the distribution within a population of the use of particular technologies, or the consumption of certain foods, can be understood as the aggregate consequences of many episodes of social learning—of episodes in which individuals learn from others to use a particular technology or to eat a certain food. The populational approach to cultural evolution began to be developed by biologists, mathematicians, and anthropologists in the 1980s (Cavalli-Sforza and Feldman, 1981) and is now pursued by a growing number of researchers from the same disciplines. Broadly speaking, these researchers fall into two groups (Clarke and Heyes,

2017; Sterelny, 2017): the "California" school, led by Robert Boyd, Peter Richerson, and Joseph Henrich, which emphasizes the importance of interactions between genetic and cultural evolution (Richerson and Boyd, 2005; Henrich, 2015); and the "Paris" school, led by Dan Sperber (Atran, 2001; Morin, 2015; Sperber, 1996), which sees cultural evolution as having greater autonomy in relation to genetic change. In contrast with the historical view, the populational conception of cultural evolution constrains cultural characteristics to those that are acquired through certain kinds of social learning (see Chapter 4). It is also more "evolutionary" than the historical view, in three respects.

First, in common with High Church evolutionary psychology, populationists assume that social learning—or, at least, the kinds of social learning that drive large-scale changes in human populations, often called "cultural learning"—is built on a set of cognitive instincts. Populationists typically assume that natural selection acting on genetic variants has given humans a set of psychological mechanisms, or "learning biases," that are specialized for learning from others. These are different from the mechanisms that enable us to learn through unassisted interaction with our environments—for example, by fiddling with a machine until we find out how it works—and, in most cases, different from the social learning mechanisms found in other animals (Henrich, 2015).

Second, populationists of the California school are very much concerned with how genetic evolution interacts with cultural change; with "dual-inheritance" or "gene-culture coevolution." This kind of coevolution occurs when a change in the socially learned characteristics of a population provokes a change in genetically inherited characteristics, or vice versa. The classic example of gene-culture coevolution is lactose tolerance (Holden and Mace, 2009). The ability to digest lactose in milk, not only in infancy but in adulthood, varies considerably across contemporary human populations. For example, it is common in Europe and Western Asia, but rare in the Far East. This distribution of lactose tolerance is thought to be due to a sequence of events in which some historical populations first started dairy farming, making milk plentifully available as a source of nutrients. This meant that the small number of people in those populations who had the genes enabling them to digest lactose in adulthood, and thereby to exploit this resource throughout their lives, had more children than others who lacked those genes. As the proportion of lactose tolerant people increased, the demand for dairy products increased, promoting dairying practices. Thus, dairying, a set of characteristics that is inherited via social learning, promoted the spread of lactose tolerance, a genetically inherited characteristic, which in turn further promoted the spread of dairying.

Third, the populationist conception of cultural evolution is evolutionary at a methodological level: it borrows techniques from the study of genetic evolution, applying mathematical models, initially developed in population genetics, to socially learned characteristics. These models are used to investigate the conditions in which cultural learning is likely to evolve, and the ways in which genetic and cultural evolution are likely to interact under different sets of assumptions. Thus, on the populationist view, cultural change is evolutionary by virtue of its interdependence with genetic evolution, because it is: (1) made possible by genetically inherited psychological mechanisms; (2) in continuous interaction with genetic evolution; and (3) subject to analysis using mathematical tools originally developed by geneticists.

The strongest conception of cultural evolution is the selectionist, or "Darwinian" approach. It assumes that cultural change is not merely interdependent with evolutionary change but that cultural change is itself evolutionary in the Darwinian sense; that the conditions required for the occurrence of Darwinian or natural selection are present in the domain of culture. These conditions were described by Donald Campbell, one of the pioneers of the selectionist approach:

> 1. A blind-variation-and-selective-retention process is fundamental to all inductive achievements, to all genuine increases in knowledge, to all increases in fit of systems to their environment.
> 2. In such processes there are three essentials: (1) mechanisms for introducing variation; (2) consistent selection processes; and (3) mechanisms for preserving and / or propagating the selected variations. (Campbell, 1974: 421)

The selectionist approach assumes that, in the cultural domain, the last of these "essentials" is met by certain kinds of social learning; the inheritance mechanisms are cultural rather than genetic. The mechanisms for introducing variation—the cultural equivalent of mutations—are generators of "error" or of "innovation" in social learning. For example, a new cultural variant could be produced by one person or group making a mistake while learning from another; trying through their own efforts to improve something acquired through social learning (using four rather than three knots to secure a fishing line, deliberately or in error); or combining information from different sources (after observing one person using three knots of type A, and another person using one knot of type B, the learner uses three knots of type B).

Describing the second requirement as a need for "selection processes" could be misleading because "selection" and "selection process" are often used to describe what happens when all three requirements are met. So, let us say that the second requirement is for "nonpurposive sorting mechanisms" (Amundson, 1989: 417). These are processes that result in some variants persisting for longer, or being copied more, than other variants, and that achieve these effects without foresight or intelligence; the sorting processes do not know which variants are more likely to be useful or adaptive. In the cultural case, these sorting processes could include intrapersonal psychological processes that make some behaviors more noticeable, learnable, or memorable than others, as well as interpersonal or intergroup processes that make people with one habit (for example, making four-knot fishing lines) more likely to survive and reproduce than those with an alternative habit (for example, making three-knot fishing lines).

Campbell's selectionist conception of cultural evolution (1965; 1974) built on Darwin's view (1871) that a variety of natural selection occurs among alternative forms of words in a language, and on Popper's proposal that scientific theories "struggle for the survival of the fittest" (1962: 68). It has been pursued by Plotkin and his students (Heyes, in press a; Plotkin and Odling-Smee, 1981; Mesoudi, 2011; Mesoudi, Whiten, and Laland, 2004)—most of whom are, intellectually if not geographically, now part of the California school—and to some degree by other members of the California school.[4]

In the late 1980s, a highly distinctive selectionist approach began to be developed by Dawkins (1989) and Dennett (1990). They coined the term "memes" to refer to cultural variants—typically understood to be ideas or beliefs—and the approach is known as "memetics" (Blackmore, 2000). Unlike other selectionist theories, memetics suggests that cultural evolution occurs "for the sake of the memes." In other theories, individual humans or social groups are the "bearers of fitness": the proliferation rate of different cultural variants is determined by processes that occur within and among people (learning, remembering, cooperating, fighting) and depends on the rate at which individuals and groups reproduce (have babies, split into two or more similar groups). In memetics, cultural variants are

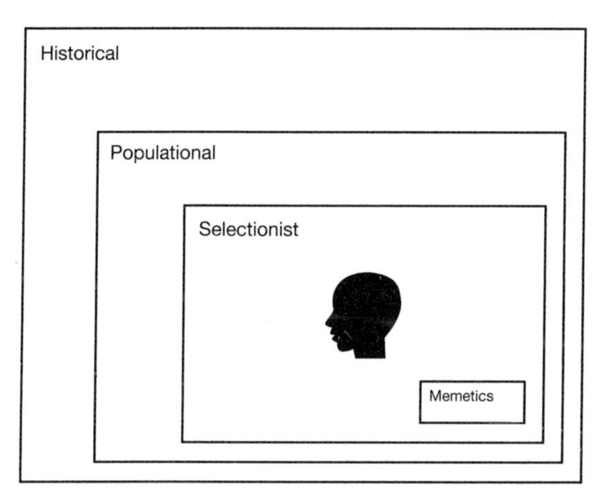

2.2 Relations between purely historical, populational, and selectionist conceptions of cultural evolution. The cognitive gadgets theory (indicated by a head icon) pursues a selectionist, but not a memetic, approach.

bearers of fitness in their own right. The spread of a meme is explained primarily by the meme's own characteristics (an idea is more or less "viral," a tune is more or less "catchy") and does not require reproduction by the individuals or groups that carry the meme (Sperber, 2000a; Sterelny, 2006).

"Meme" is a very successful meme. The word itself, and the central idea—that ideas are agents competing to get copies of themselves into as many minds as possible—has captured the public imagination. However, most scientists and philosophers doubt that memetics is sufficiently coherent, or grounded in empirical evidence, to provide the basis for a research program (Lewens, 2015).

Figure 2.2 shows the relationships among the various conceptions of cultural evolution. The populational approach is nested within the historical approach; it takes cultural evolution to be evolutionary, not only because it involves change over time, but also because cultural change interacts with genetic evolution and can be analyzed using similar mathematical tools. The selectionist view is nested within the populational approach. Its distinctive, evolutionary claim is that the conditions for natural or Darwinian selection—variation-and-selective-retention—hold in the cultural domain.

As it happens, many contemporary selectionists—including some who work in memetics—assume that cultural learning is built on genetic adaptations, regard gene-culture coevolution as normative, and use mathematical models from population genetics to analyze cultural change. However, the selectionist view does not entail a close or continuous causal relationship between genetic and cultural evolution. In this book, I pursue a "Campbellian" selectionist approach (Campbell, 1965; 1974). This assumes that genetic and cultural evolution are based on the same, fundamental heuristic—variation-and-selective-retention—and embrace the populational assumption that large-scale cultural change is the aggregate consequence of many episodes of social learning. However, it suggests that, rather than being on a short "genetic leash" (Lumsden and Wilson, 2005: 144), cultural evolution is highly autonomous with respect to genetic evolution. Culture has many of its most profound effects without influencing or being influenced by genetic evolution.

CULTURAL EVOLUTION OF COGNITIVE MECHANISMS

Cultural evolutionary theory has been used primarily to explain changes over time in behavior (for example, the skills involved in making tools, or the practices that constitute dairying) and artifacts (for example, the designs of stone tools, boats, and fish hooks).[5] It has also been applied to large-scale conceptual structures such as fairy tales, systems of religious belief, and scientific theories. All of these explanatory targets—behavior, artifacts, and conceptual structures—are types of cognitive grist; they are components of the external, social world that are "taken in" (and produced) by human minds. Cultural evolutionary psychology extends the analysis from grist to mills—to the internal cognitive processes that grind the grist. It suggests that cultural evolutionary theory can illuminate not only what the mind works on, but how the mind works—the way in which cognitive processes operate.

The populational view may be sufficient to provide some insight into the origins of distinctively human cognitive mechanisms, but the selectionist approach has greater potential. When the populational assumption (that large-scale cultural changes are the aggregate consequences of many episodes of social learning) is combined with the selectionist view (that cultural change can be driven by variation-and-selective-retention), cultural evolution has the potential to explain the adaptedness of distinctively human cognitive mechanisms; why, in some cases, cognitive mechanisms seem to fit the environments in which they operate and do their jobs reasonably well. However, the selectionist approach is demanding. To apply it in a new domain, such as the domain of cognitive mechanisms, we need to formulate clear hypotheses about: (1) the entities that are evolving; (2) the routes of their inheritance; and (3) the kinds of social learning that provide an inheritance system.

Variants

How can a stream of behavior be divided into discrete chunks? For a behavior to be "new," by how much, and in what way, must it be different from the "old" behavior? These "unitization" or "individuation" questions are important because in order for change to be explicable by Darwinian selection, whether genetic or cultural, the thing-that-is-changing must be quantifiable. However, it is very hard to answer in a principled and rigorous way questions about the unitization of psychological grist—ideas and behavior (Brown, 2014). Arguably, this is the reason why, after thirty years of conceptual development, memetics still has not been converted into an empirical research program, and its hypotheses still rarely inspire observation and experiment.

Unitization problems are more tractable for cognitive mechanisms than for the traditional targets of cultural evolutionary analysis. Ideas and behaviors (grist) are unitized by common sense or folk psychology, which is not designed for scientific thinking in general or for evolutionary analysis in particular. If I think about heaven, am I having just one thought, or one thought for every cloud and angel in my mental picture? If I walk to the shop, am I doing one thing or many things—exercising my muscles, moving my legs, wearing clothes, and carrying a shopping bag? Folk psychology does not answer these questions, or tell us how they can be answered. In contrast, cognitive mechanisms (mills) are unitized by cognitive science, and, within the framework of cognitive science, different versions of a mechanism—different variants—can be distinguished according to what they do and how they do it; the kind of information they process, and the computations they use to process that information (Heyes, 2003).

For example, a "reading aloud mechanism" (such as those shown as "routes" in Figure 1.3) is capable of converting script into speech. One reading aloud mechanism can differ from another in terms of the range of script sequences it can convert into speech (for example, only words that are pronounced according to certain rules, or both regular and irregular words). Two reading aloud mechanisms can also differ according to the kinds of representations they process and the computations that do the processing. The "grapheme-phoneme" reading aloud mechanism works with sensory representations—shadowy mental images of the look and the sound of word components—

whereas the "lexical semantic" mechanism also involves abstract, structured representations of the world (Coltheart et al., 2001). Furthermore, whereas folk psychology allows my head to contain indefinitely many versions of the same idea (for example, umpteen different mental images of "an angel" or "a cloud"), the framework of cognitive science stipulates that each head contains just one token of any given type of cognitive mechanism. My brain might contain a grapheme-phoneme mechanism as well as a lexical semantic mechanism for reading aloud in English, but, by definition, it does not contain two slightly different grapheme-phoneme mechanisms.

Thus, cultural evolutionary psychology suggests that distinctively human cognitive mechanisms are entities subject to Darwinian cultural evolution. Cognitive science, rather than folk psychology, unitizes cognitive mechanisms. In cognitive science, variant cognitive mechanisms are defined in a functional way; the traits, or characteristics, of a cognitive mechanism relate to the kind of information it can process, and the computations and representations it uses to process this information.

Routes of Inheritance

Genetic inheritance occurs between biological parents and their offspring, but, in principle, any member of a population can culturally inherit information from any other member. Three routes of cultural inheritance have been distinguished: vertical, oblique, and horizontal (Cavalli-Sforza and Feldman, 1981). Vertical inheritance is from biological parents to their offspring; oblique inheritance is from individuals of one biological generation to genetically unrelated or distantly related individuals of the next biological generation (for example, from school teachers, aunts, and uncles); and horizontal inheritance is between individuals of the same biological generation (for example, friends, co-workers, and siblings). Any given cultural trait can be inherited via one or any combination of these routes, and the importance of each route is likely to vary, not only with the nature of the trait, but also across cultures. For example, in contemporary Western societies, religious beliefs may be more likely than technological skills to be vertically inherited, and culinary skills may be predominantly vertically inherited in one culture but predominantly obliquely inherited in another.

Like beliefs and skills (grist), cognitive mechanisms (mills) may vary among themselves, and across cultures, in the extent to which cultural inheritance depends on the vertical, oblique, and horizontal routes. However, a "distributed" pattern is likely to be common, in which all three routes see heavy traffic, albeit at different times in development. Take mindreading as an example (Chapter 7). In contemporary Western societies, the foundations of the capacity to ascribe thoughts and feelings are likely to be laid in infancy and early childhood through interaction with parents (the vertical route) and other members of the parental generation (the oblique route; see Lewis, Freeman, Kyriakidou, Maridaki-Kassotaki, and Berridge, 1996). However, in school-age years, when children talk to one another about mental states, and in adolescence and adulthood, when many people both continue "live" discussion of the mind and read literary fiction (Kidd and Castano, 2013), the development of mindreading is influenced predominantly by peers (horizontal) and older, genetically unrelated individuals (oblique).

The distributed pattern is likely to be especially common for social cognitive mechanisms, like mindreading, because these mechanisms are not only acquired through social interaction, but used and broadcast in social interaction throughout the lifespan. For example, in cultures where mindreading is valued (Duranti, 2008), adults broadcast the workings of their theory of mind when they comment on the thoughts and motivations of others. Because mindreading has a regulatory as well as a predictive function (McGeer, 2007), these broadcasts are likely to fine-tune listeners' theories of mind, and thereby to increase the similarity between the theories held by members of the same social group. In contrast, causal understanding—a mechanism that represents certain kinds of interactions between physical objects—may depend on social interaction for its development, but causal understanding is not used or broadcast as intensively as mindreading in social interaction throughout life. Therefore, I could have a highly distinctive mechanism of causal understanding, culturally inherited from my parents or teachers, which is not passed on to the people I deal with every day, because the way it works is invisible to them; I rarely use it in company, and, when I do, it is without commentary.

Distributed inheritance tends to reduce within-group variation. The members of each group become increasingly similar to one another, and increasingly different from the members of other groups. Therefore, when a type of cognitive gadget is culturally inherited in a distributed way, and when the success of the gadget is dependent on its social context, it is possible to get an adaptive response at the level of the social group. Although there is only one token gadget in each head, the fittest gadgets would be those that are most effective in furthering the projects of the social group. The fitness of a variant gadget can be measured by comparing the number of groups in which all or most heads carry the focal variant, with the number of groups in which all or most heads carry an alternative variant (or just the relative proportions of each variant across the whole population's heads—see Chapter 9). For example, although each head contains only one theory of mind, the success of an "honor theory" of mind—a theory in which the desire to retaliate against insults is an important source of motivation—can be measured by comparing (other things being equal) the number of groups in which all or most people carry the honor theory, with the number of groups in which all or most people carry a theory of mind in which honor plays little or no part.

Predominantly vertical inheritance may occur when the development of a cognitive mechanism is almost wholly dependent on social interaction in infancy, and in cultures where infants interact almost exclusively with their biological parents. In principle, the fitness of a vertically inherited cognitive gadget could be measured by comparing the number of biological offspring, or "babies," produced by people who have a focal variant, with the number of babies produced by people who have an alternative variant. In other words, the cultural fitness of the cognitive gadget would correspond with the biological, reproductive fitness of its bearers, and individuals, rather than social groups, would be the adapted systems.

Many specialized skills (grist) are culturally inherited in an oblique pattern. For example, in medieval Europe, craft skills—such as glassmaking, stonecutting, and cordwaining—were culturally inherited via master-apprentice relationships, in which the master was a member of the same generation as the apprentice's biological parents, but not from the same family. This kind of cultural inheritance is not of focal interest here because, following evolutionary psychology more generally, cultural evolutionary psychology is concerned primarily with the origins of distinctively human thinking; with cognitive processes that are present, in one form or another, in all or most adults alive today, and absent, or present only in nascent form, in nonhuman animals. It focuses on the cognitive processes—such as mental mapping, mindreading, and imitation—that make us such peculiar animals; that constitute human nature (Samuels, 2012; see Chapter 9). Variant forms of these types of cognitive process—for example, slightly different ways of representing space, and different theories of mind—play a very important part in the story. As Campbell's first requirement reminds us, variation is one of the three "essentials" for evolution by natural selection. But types of cognition that are found only in a small minority of humans, such as the type of cognition required for cordwaining or calculus, are not in the core domain of cultural evolutionary psychology. These minority types of human cognition would enter the core domain only if, like literacy in recent decades, there was a rapid increase in their prevalence across the global population.

Mechanisms of Inheritance

Inheritance mechanisms are a key requirement for Darwinian selection. There have to be mechanisms "for preserving and / or propagating the selected variations" (Campbell, 1974); mechanisms that make offspring similar to their parents. In the case of genetic evolution, these are primarily the mechanisms of DNA replication. In the case of cultural evolution, they are typically said to be mechanisms of social learning (Boyd and Richerson, 1985; Richerson and Boyd, 2005). Episodes of social learning, in which one person learns behavior or ideas (grist) from another, are ubiquitous in everyday life—I learn to avoid eating certain mushrooms by noticing that you avoid them; you show me how to knead dough to make bread; or you tell me about the Higgs boson. Describing such episodes as examples of "copying" or "transmission" can give the misleading impression that social learning is exactly analogous to genetic replication, but it is now widely recognized that social learning is seldom, if ever, a process in which pellets of information are duplicated and moved from one place to another. Certain kinds of social learning, circumscribed as "cultural learning," allow ideas and behavior to be inherited with sufficient fidelity for selection to operate, but they do not fulfil this inheritance function in the same way as the mechanisms of DNA replication. Rather, as we shall see in Chapters 4–8, the psychological mechanisms that allow one person to learn an idea or skill from another, and to preserve it long enough to pass it on to a third party, are often complex and indirect (Heyes, 1993; Morin, 2015; Sperber, 1996).

This insight opens up the possibility that features of cognitive mechanisms (mills) are culturally inherited in the same ways as ideas and behaviors (grist). A cognitive mechanism certainly is not a pellet of information that can be copied inside your head, sent through the air, and planted wholesale in my head. But if grist is not "copied" in this sense, there is no reason to expect mills to be "copied," either. Instead, we can recognize that certain kinds of social interaction, sometimes with many agents over a protracted period of time, gradually shape a child's cognitive mechanisms so that they resemble those of the people around them. Reading, discussed in Chapter 1, is a clear example. Everyone acknowledges that children are typically taught to read, that literacy training produces new neurocognitive mechanisms, and that we do not genetically inherit any specific predispositions to develop these mechanisms. Cultural evolutionary psychology merely draws attention to the fact that literacy training is a set of social interactions—providing demonstrations, instructions, feedback, and encouragement, in formal and informal settings—and that these social interactions result not only in the inheritance of skills (grist), but in the inheritance of cognitive mechanisms (mills). If literacy training were achieved by planting a "reading chip" in each child's brain, the cultural inheritance of reading would be more like the genetic inheritance of eye-color, but it would not necessarily be more effective as an inheritance mechanism, that is, in preserving and / or propagating selected variants (Campbell, 1974).

NATURE, NURTURE, CULTURE—IN PRACTICE

The teleosemantic view provides a way of comparing inherited and environmental contributions to development in a common currency of information. Research on cultural evolution suggests that, when it comes to explaining human development, we need to think not only about genetically inherited information, but also about culturally inherited information. Traditionally, genetic contributions to development are said to be "nature," and environmental contributions, to be "nurture." Adopting that convention, "culture"—or culturally inherited information—is a third kind of contribution to development that possesses some features of nature, and some of nurture. Culture is "nature-like" as a product of a selection-based inheritance system, but "nurture-like" in being acquired in the course of development through interaction with the (social) environment.

In practice, how can we tease apart the contributions of nature, nurture, and culture to the development of cognitive mechanisms? It is certainly not easy, but Chomsky (1965) provided an important clue when he argued that language learning must be guided by genetically inherited knowledge of grammar, a "language acquisition device," because there is "poverty of the stimulus." In other words, children could not learn grammar exclusively from their environments because those environments do not provide enough information of the right kinds. The information provided by what others say, and by feedback on the child's efforts to speak, is too incomplete, or "poor," to explain how children master the grammar of their native languages. The merits of this particular argument, with respect to a genetically inherited language acquisition device, will be discussed in Chapter 8. Of interest here is the fact that Chomsky's argument can be generalized in two ways: from language to other cognitive mechanisms, and from poverty alone to a contrast between poverty and "wealth" of the stimulus (Ray and Heyes, 2011). Thus, the development of any cognitive mechanism is characterized by poverty or wealth of the stimulus according to whether the environment of development provides too little (poverty) or at least enough (wealth) usable information to explain the properties of the cognitive mechanism.

Broadly speaking, evidence of poverty and wealth comes from patterns of covariation. Poverty of the stimulus is indicated when the development of a cognitive ability does not vary reliably with environmental factors, and wealth of the stimulus is indicated by strong covariation with environmental factors. Poverty is a sign that the development of an adaptive cognitive trait is dependent on genetically inherited information (nature). Wealth is a sign that development is dependent on learning in a broad sense (nurture) and / or culturally inherited information (culture). Nurture is indicated when cognitive development varies with features of the environment in which development is actually occurring; with information that can be acquired by "unassisted" or "asocial learning," and by the kinds of social learning found in a broad range of animals. Culture is indicated when cognitive development varies with longer-term features of the environment; features that may not be present when a particular individual is developing, or that can be acquired only via certain kinds of social learning, known as cultural learning. (The relationships between learning, social learning, and cultural learning are discussed further in Chapter 4.)

Variation in cognitive ability can be measured across: (1) time points in development, (2) groups or individuals within a human population, (3) human populations, or (4) species. An ideal study of the first kind of variation would test a cognitive ability (for example, counting) in the same group of children repeatedly over time, and, between tests, record the opportunities these children have in their daily lives to acquire, through interaction with objects and other people, the information needed to increase the cognitive ability. If the opportunities predict the increases in cognitive ability over time, this kind of study provides evidence of wealth of the stimulus. If the opportunities are a poor predictor of cognitive development—for example, if changes in the ability to count occur in fits and starts, or outstrip opportunities for learning—it provides evidence of poverty of the stimulus. A common variant, or limiting case, of this kind of study simply tests for a cognitive ability in newborn babies. These studies typically assume that newborns have had little or no opportunity for learning and, therefore, that any cognitive ability they possess indicates poverty of the stimulus.

Studies examining variation within a human population usually test a cognitive ability at one time-point and ask whether differences in ability between groups, or among individuals, are related to the opportunities for relevant learning before the test. For example, are children whose mothers have talked to them often about thoughts and feelings—who have had many opportunities for cultural learning about the mind—better able to infer beliefs, or to recognize emotions, than children whose mothers have talked to them relatively little about mental states (Taumoepeau and Ruffman, 2006; 2008)? If the answer is no, we have evidence of poverty, and if the answer is yes, we have evidence of wealth. The "twin studies" used in behavioral genetics fall into this category. In a typical twin study, a cognitive ability is measured at one time-point in a large number of children who are either monozygotic (identical) or dizygotic (fraternal) twins. Using the fact that monozygotic twins have identical genotypes, whereas dizygotic twins have an average of 50 percent of their genes in common, these studies compare variation in cognitive ability within monozygotic pairs with variation within dizygotic pairs to calculate the extent to which development of the cognitive trait is based on genetically inherited information, rather than information derived from the environment (Plomin, DeFries, McClearn, and McGuffin, 2001). A weak environmental contribution indicates poverty of the stimulus, and a strong environmental component is evidence of wealth.

Research examining variation across human populations is the province of cross-cultural psychology. Cross-cultural studies typically compare samples of children or adults from broadly defined populations—for example, "Western Caucasians" versus "East Asians"—that are known to give children different kinds of experience as they grow up. If the nature of this experience covaries with cognitive characteristics—for example, if growing up in an "individualistic" society is associated with elemental processing of visual scenes, whereas growing up in a "collectivist" society is associated with holistic processing (Senzaki, Masuda, and Ishii, 2014)—there is evidence of wealth of the stimulus; and to the extent that the environmental and cognitive variations are unrelated, there is evidence of poverty.

Finally, research involving nonhuman animals—in comparative psychology, ethology, and primatology—can provide important evidence relating to the poverty or wealth of the stimulus and, more generally, help us to tease apart the contributions of nature, nurture, and culture to human cognition. It is widely agreed that, although some species of nonhuman animals have "traditions"—they show local variations in behavior due to social learning—nonhuman animals do not have "culture," as culture is characterized by selectionist theories. Specifically, in nonhuman animals, modifications to socially learned characteristics cannot accumulate over time in a way that affords improvement; that allows beliefs, artifacts, practices (grist)—or, indeed, cognitive mechanisms (mills)—to get better at doing their jobs (Caldwell, Atkinson, and Renner, 2016; Tomasello, Kruger, and

Ratner, 1993). Consequently, at the most basic level, if a cognitive ability is found not only in humans but also in other animals, its development is very unlikely to depend on culture. More generally, when species that are closely genetically related to humans, such as chimpanzees, have a more human-like cognitive capacity than species that are distantly related to humans, such as rats, all other things being equal, it suggests that development of the focal capacity is heavily dependent on genetic information. In contrast, to the extent that variations in cognitive capacity across species correlate not with genetic relatedness but with ecological factors, and therefore with opportunities for learning, it suggests either that there has been convergent evolution, or that development of the cognitive capacity is highly dependent on learning. Bat wings and bird wings are a morphological example of convergent evolution. They are similar in structure, and the development of both kinds of wings is highly dependent on genetic information, but they have different evolutionary histories. Convergent evolution can be distinguished from learning-dependent development through studies of one or more of the other kinds of cognitive variation; over time, across individuals or groups within a population, and across populations.

Each of these four sources of evidence relating to poverty and wealth of the stimulus depends exclusively on correlating differences in cognitive ability with spontaneous variation in opportunities for learning and social learning. However, experimentally induced environmental variation, "training," can also be a powerful research tool. For example, hypotheses about the kinds of learning that contribute to cognitive development can be tested by giving some individuals—children, adults, or nonhuman animals—more or fewer opportunities for learning between successive cognitive tests, or prior to a single test (Lohmann and Tomasello, 2003). Provided the training is not too artificial—that it exposes individuals to environments similar to those in which the focal cognitive ability normally develops—wealth of the stimulus is indicated when training has an impact on cognitive development, whereas poverty of the stimulus is indicated when it does not.

It is far from easy to parse cognitive development—to identify the contributions of nature, nurture, and culture to the formation of a cognitive mechanism—and each of the methods outlined above is highly fallible. When learning opportunity A (for example, talking with a parent about mental states) correlates with cognitive ability B (mindreading), it could be because a hidden factor C (linguistic skill) is influencing both A and B, not because A is causing B. Likewise, twin studies may indicate a relatively large genetic contribution to development simply because the people included in the study happen to have grown up in very similar environments, and, in cross-species comparisons, convergent evolution can be mistaken for a strong influence of learning on development. Given these risks, in this area of science, as in most others, we have to place more trust in research that includes effective control procedures, and to look for convergent evidence—for signs that studies using different samples and methods are pointing to the same conclusion.

CONCLUSION

In summary: the teleosemantic view suggests that a biological structure X carries information about Y only if the state of X correlates with Y, *and* X was selected because its states correlate with Y. This conception of information provides a theoretical framework in which to isolate and compare the contributions of nature (genetically inherited information), nurture (information derived from direct interaction between the developing system and its environment), and culture (culturally inherited information) to human cognitive development.

Populational models of cultural evolution suggest that culture is a major contributor to the development of cognitive grist—distinctively human behavior, artifacts, and conceptual structures. I argue that a specific kind of populational model—a selectionist approach—can also be used to understand the development of cognitive mills—distinctively human cognitive mechanisms. The selectionist approach suggests that the conditions necessary for Darwinian evolution—variation, selection (or "sorting"), and inheritance—are present in the cultural domain. Starting the process of applying a selectionist analysis to cognitive mechanisms, I have formulated hypotheses about variants, routes of inheritance, and mechanisms of inheritance, and discussed the kinds of empirical evidence that can help us isolate the roles played by cultural inheritance in the development of cognitive processes.

This chapter started with an acknowledgement that the development of every aspect of human behavior and cognition depends on a rich stew of factors. I have argued that, in spite of this multifactorial complexity, the contributions of nature, nurture, and culture to the development of any given human cognitive trait can and should be identified with the help of the teleosemantic conception of information, combined with empirical studies investigating poverty and wealth of the stimulus. In the next chapter, I use some of these tools to identify the "starter kit" of human cognition: a set of genetically inherited psychological characteristics that make a major contribution to the development of distinctively human cognition.

3

STARTER KIT

THE MIND OF A NEWBORN BABY IS *NOT* A TABULA rasa, and the marks on the human slate at birth are subtly but crucially different from those on the chimpanzee slate. At sixteen weeks of gestation, human brains are already twice as large as chimpanzee brains (Rilling, 2014). This alone suggests that we humans have an extensive "genetic starter kit."[1] Much of this book emphasizes the importance of learning and cultural inheritance in the development of distinctively human cognition, but this chapter focuses on genetic inheritance. It concerns the inborn temperamental factors, attentional biases, and central cognitive processes that help to make mature human minds so very different from those of other animals.

High Church evolutionary psychology and cultural evolutionary theory suggest that humans genetically inherit "Big Special" psychological attributes: programs or blueprints for the development of powerful cognitive mechanisms that are qualitatively different from those found in other animals, such as language, theory of mind, imitation, causal understanding, episodic memory, self-conception, and face processing. In contrast, this book suggests that, although the vast majority of adult humans have these Big Special cognitive mechanisms, we do not genetically inherit programs for their development. Rather, we genetically inherit "Small Ordinary" psychological attributes: the propensity to develop relatively simple mechanisms that closely resemble those found in other animals, including chimpanzees. Genetic evolution has tweaked the human mind.

The genetically inherited differences between our minds and those of our ancestors are small but very important. They enable the development of Big Special cognitive mechanisms in three ways. First, genetically inherited changes in *temperament* have helped to make humans remarkably social primates. Compared with chimpanzees, we tolerate, seek, and thrive on the company of other agents. This constitutional sociality greatly increases children's opportunities to acquire not only knowledge and skills (Henrich, 2015; Sterelny, 2012) but also cognitive mechanisms from other agents in the course of development. Second, genetically inherited *attentional biases* ensure that the attention of human infants is locked-on to other agents from birth. We are driven from very early infancy to look at biological motion and faces, and to listen to human voices. Whether or not these tendencies evolved specifically for pedagogy (Csibra and Gergely, 2006; Heyes, 2016b), their genetic inheritance means that, as soon as human babies emerge into the world, they start extracting information as well as care from the adults around them. Finally, humans have uniquely powerful *central processors*: mechanisms of learning, memory, and control that extract, filter, store, and use information. Each of these processors is domain-general, crunching data from all input domains using the same set of computations, and taxon-general, present in a wide range of animal species. However, humans genetically inherit central processors with unprecedented speed and capacity. Shaped and fed throughout development by the torrents of culturally evolved information flowing in from other agents, domain-general central processors not only capture this information but use it to build new, domain-specific cognitive mechanisms—the Big Special mechanisms that make humans such peculiar animals. Thus, the Big Special mechanisms are designed by cultural evolution, but they are built in the course of development by souped-up, genetically inherited mechanisms of learning and memory, using raw materials that are, from birth, channeled into infant minds by genetically inherited temperamental and attentional biases.[2]

Subsequent chapters, especially Chapter 6 on imitation, look at the construction process that explains how one set of genetically inherited domain-general cognitive mechanisms could produce another set of domain-specific cognitive mechanisms using culturally inherited information. This chapter focuses on components of the genetic starter kit, surveying some of the psychological characteristics that make a major contribution to the development of distinctively human cognition, and are likely to be genetically inherited.

EMOTION AND MOTIVATION

Social Tolerance

Compared with chimpanzees, adult humans, especially males, are remarkably tolerant of one another and towards juveniles (Burkart, Hrdy, and van Schaik, 2009; Silk and House, 2011). As Susan Hrdy has pointed out, people crammed together on an airplane disguise their irritation and get along; by comparison, in a planeload of chimpanzees, "Bloody earlobes and other appendages would litter the aisles" (Hrdy, 2011: 3). No doubt much of this social tolerance, or lack of aggression, is a consequence of human enculturation, but there is evidence that a tendency to be docile rather than irascible—to accept the presence and activities of others without protest—is part of our genetic starter kit.

For example, an intriguing study found archaeological evidence that human heads and faces have changed over the last two hundred thousand years: there has been a reduction in the average brow ridge projection and a shortening of the upper facial skeleton (Cieri et al., 2014). Describing these changes as "craniofacial feminization," Cieri and colleagues argued that they are due to a reduction in androgen reactivity (lower levels of adult-circulating testosterone or reduced androgen receptor densities) that was favored by natural selection operating on genetic variants because it reduced aggression among members of our species. The link with androgen reactivity is supported by signs that androgen is involved in making bonobos more socially tolerant than chimpanzees (Wobber, Hare, Lipson, Wrangham, and Ellison, 2013). The suggestion that the recent changes in human craniofacial morphology were genetically inherited is supported by studies, initiated by Darwin (1868), of domestication in a range of nonhuman species, including wolves. These studies show that selection for docile acceptance of captivity and human contact, operating on genetic variance, rapidly produces both the targeted forms of social tolerance and a syndrome of other traits including changes in craniofacial morphology like those studied by Cieri and colleagues (Wilkins, Wrangham, and Fitch, 2014). The craniofacial components of the domestication syndrome are typically regarded as by-products of selection, but, in the hominin line, they could also have been targets of selection; individuals with lower androgen reactivity are less likely to initiate aggression but also (because they have more juvenile, rather than more "feminine," faces) less likely to elicit aggression from others.

Increased social tolerance may or may not have evolved for cultural inheritance. It is possible, as Cieri and colleagues suggest, that humans became more socially tolerant recently—about eighty thousand years ago, with the onset of behavioral modernity—and that socially tolerant individuals had more biological offspring because they were better able to acquire knowledge and skills via social learning. But it is also possible that social tolerance began to increase much earlier in the hominin line and was selected because it gave access to other advantages of group living, such as security and food sharing. Whatever the selective history, it is clear that the high levels of social tolerance exhibited by most human adults, and primed by our genetic starter kit, enable human infants and children to learn much more, and much better, from conspecifics than their chimpanzee counterparts. At the simplest level, high levels of social tolerance enable human juveniles to observe and interact with a wide range of expert models, and to get close enough to each model, for a long enough period, to allow the observer to encode details of the model's skilled actions and of the contexts in which they occur (Coussi-Korbel and Fragaszy, 1995; van Schaik and Pradhan, 2003). In addition, high levels of social tolerance allow adults to become effective teachers, that is, to be patient enough to go slowly when their actions are being observed, and to stop or correct pupils' errors without evident aggression. The evolution of human teaching—whether genetic or cultural (Heyes, 2016b)—involved a good deal more than just promoting patience, but high levels of social tolerance are a fundamental prerequisite for effective tuition.

Social Motivation

Another plausible candidate for the genetic starter kit is enhanced social motivation. Compared with other primates, including nonhuman apes, it is likely that humans have a stronger inborn tendency not only to be less aggressive towards conspecifics (social tolerance) but actively to seek social contact and find it especially rewarding (social motivation). This is suggested by evidence that young children are much more likely than young chimpanzees to choose collective over solitary activities (Tomasello, 2014), combined with studies showing that, in humans and other animals, individual differences in social motivation are genetically heritable (Skuse and Gallagher, 2011).

Much recent research on social motivation has focused on neuropeptides, particularly oxytocin (Gangestad, 2016). It has been reported that the administration of oxytocin, by intravenous injection or via a nasal spray, leads people to judge the emotional facial expressions of others as especially intense (Cardoso, Ellenbogen, and Linnen, 2014), to show more trust and generosity when playing economic games (Kosfeld, Heinrichs, Zak, Fischbacher, and Fehr, 2005; but see also Lane et al., 2015), and to experience increased activity in neural reward networks when looking at images of infants or erotic stimuli (Gregory, Cheng, Rupp, Sengelarb, and Heiman, 2015). Research on oxytocin, once hailed as the "love molecule" (Yong, 2012), is now yielding a more complex and interesting picture. The effects of oxytocin administration on human behavior are often gender- and context-specific (Scheele et al., 2013), and, in addition to positive and negative changes in social attitudes, they include appetite suppression (Lawson et al., 2015) and anxiety reduction (Churchland and Winkelmann, 2012). Consequently, it is not yet clear whether oxytocin has its prosocial effects by increasing the salience of social stimuli, amplifying the anticipated and experienced value of social rewards, decreasing fear of others, or by involving, in some measure, all of these routes. What is clear is that any and all of these changes in social motivation would increase the power of other minds to shape cognitive development. Like enhanced social tolerance, greater social motivation gives juveniles better access to adult models for social learning, and equips adults to teach (Goldstein and Schwade, 2008). Increased social motivation also makes minds more malleable by the social environment. Highly attentive to the actions of others (see next section), and craving social approval, developing humans are inclined to adopt those actions, beliefs, and ways of thinking that yield social rewards—nods, smiles, warm touches, kind words—and, as every parent knows, the social rewards need not be delivered deliberately, with the intention of teaching or improving behavior, to have an impact. Anything that pleases significant others—because it is funny, useful, or familiar in adult behavior—is liable to be incorporated into the child's behavioral repertoire and view of the world.

It has been suggested that the key to human social motivation lies in our enjoyment of "response-contingent stimulation," that is, of events that are predicted by, and occur shortly after, our own actions (Csibra and Gergely, 2006; Heyes, 2016b). In principle, response-contingent stimulation can be social or inanimate. Indeed, some of the earliest evidence that human infants enjoy response-contingent stimulation came from studies showing that infants smile, laugh, and coo when looking at a mobile—an inanimate stimulus—when the mobile had, in the past, rotated whenever the infants turned their heads (Watson and Ramey, 1972). We like making things happen, whether the things are social or asocial. However, in everyday life, it is often social things that we are able to control, and therefore the reactions of other agents are a major source of response-contingent stimulation.

Evidence that our enjoyment of contingencies is genetically inherited comes from a study showing that, in human newborns, actions that are followed by response-contingent stimulation increase in frequency (they are subject to reinforcement learning) not only when the stimulation is biologically relevant—for example, the delivery of milk—but also when it consists of short, emotionally neutral speech sounds (Floccia, Christophe, and Bertoncini, 1997). This suggests that, for human newborns, some stimuli are attractive not by virtue of their intrinsic, biological properties (primary reinforcers) or because they have been paired with primary reinforcers (secondary reinforcers) but simply because they have been experienced in a contingent relation with the infant's responses.

Humans are certainly not the only animals that enjoy response-contingent stimulation. Rats snuggle up to a human hand that has been contingently responsive (Werner and Latane, 1974), and bob-white quail chicks approach the sound of a maternal call that has been heard in a contingent, rather than a non-contingent, relationship with the chick's own distress calls (Harshaw and Lickliter, 2007). However, it is eminently plausible that genetic evolution has tweaked this tendency, making humans more inclined than our primate ancestors to enjoy response-contingent stimulation, and therefore to approach and learn from interaction with other agents.

Enhanced social motivation would have a major impact on development whether or not the enhancement is underwritten by oxytocin, but research on this neuropeptide emphasizes the fact that Small Ordinary changes can have major downstream effects. Oxytocin is part of an ancient peptide signaling system found in nearly all animals, including insects and worms (Gruber, 2014). Originating about 700 million years ago, in various taxa, this system has been implicated in the control of muscles used in reproduction (for example, uterine contractions and egg laying), in ejaculation and sperm release, in the ejection of milk from mammary glands, and in infant care, consortships, and pair-bonding (Gangestad, 2016). Thus, oxytocin is part of a system that has been repeatedly tweaked by genetic evolution to fulfil new functions. For example, in vole species, the transition from primarily promiscuous to primarily monogamous mating systems involved the insertion of an oxytocin (or vasopressin) receptor gate in a basic mechanism of reinforcement learning (Insel and Young, 2001). Therefore, any changes that occurred in the hominin line are likely to have been further tweaks or small adjustments to exactly how the system works in concert with others, and in different stimulus environments, conserving and extending old functions (Anderson and Finlay, 2014). In humans, upregulation of oxytocin may have incrementally extended social motivation—the salience of social stimuli, the value of social rewards and/or the tendency to view others as "safe." In themselves, these are also small, genetically inherited effects. But because the small effects make developing minds more malleable by the social environment, they allow cultural evolution to do much more than tweaking. They allow cultural evolution to fill developing minds with knowledge and skills, and to build Big Special cognitive mechanisms.

ATTENTION

Social tolerance and motivation get developing humans up close and personal with a wide range of adults; adults who are equipped to fill and shape their minds with culturally inherited information. Input biases ensure that, from birth, human children target their attention on these experts, ready to drink in the information they have to offer (Heyes, 2003).

Faces

Humans share with other species an inborn preference for biological motion; like freshly hatched chicks, newborn babies look longer at the movements of an animal (human or nonhuman) than at the movements of inanimate objects (Bardi, Regolin, and Simion, 2011; 2014). This genetically inherited input bias, which is thought to be evolutionarily ancient, trains attention on living things. That is a good start, but studies of newborns indicate that humans also genetically inherit a more specific input bias making them selectively attend to the body-part where the most information is available: the face. Compelling evidence of an inborn face preference comes from studies

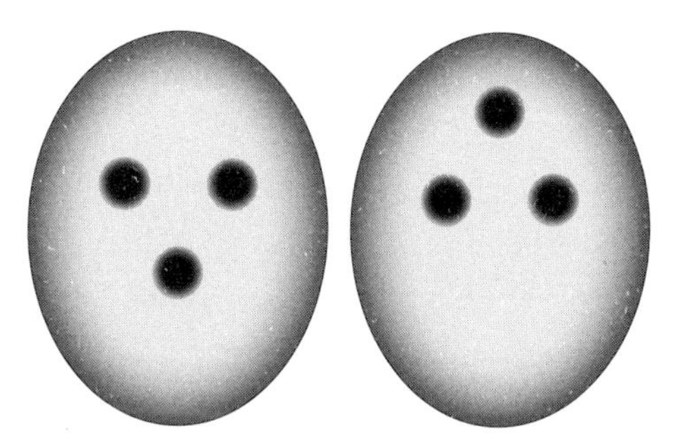

3.1 Stimuli of the kind used to test for an attentional bias towards faces in human newborns. (Reprinted with permission from Johnson, 2005.)

showing that human newborns turn their heads for longer to keep looking at a triangle of three dark blobs with the apex at the bottom (left side of Figure 3.1)—a face-like stimulus—than at an array which, although it has the same low-level visual features, is not face-like, because the apex of the triangle is at the top (right side of Figure 3.1; Johnson, 2005; Johnson, Dziurawiec, Ellis, and Morton, 1991). A recent study involving fetuses in the third trimester of pregnancy confirms that the human face preference does not depend on experience with faces. When stimuli like those in Figure 3.1 were projected through mothers' abdomens, fetuses turned their heads more often towards the upright than the inverted triangle of blobs (Reid et al., 2017). Something akin to an inborn face preference has been found in newly hatched chicks (Rosa-Salva, Regolin, and Vallortigara, 2010), but not in chimpanzees or other nonhuman apes (Kano, Call, and Tomonga, 2012). Therefore, it is possible that convergent genetic evolution is responsible for the face preference in humans and precocial birds, and the human face preference emerged or was amplified in the hominin line.

The inborn face preference is a simple thing, triggered by triangles of dark blobs, but this genetically inherited input bias provides a foundation for the development of a whole set of increasingly specific and complex attentional processes. At birth, the preference is not specific to human faces. Newborns can distinguish human and monkey faces but do not prefer the former over the latter (Di Giorgio, Leo, Pascalis, and Simion, 2012). However, at three months of age, infants look longer at chimpanzee faces containing human eyes than at chimpanzee faces containing chimpanzee eyes (Dupierrix et al., 2014); at four months, they prefer human faces making direct eye contact to faces with averted gaze (Vecera and Johnson, 1995); and by six to nine months, the face preference is not only human-specific (Pascalis, de Haan, and Nelson, 2002) but strong enough to overwhelm the attentional pull of a wide range of other, highly salient, dynamic stimuli (Frank, Vul, and Johnson, 2009). Yet, there is no reason to assume that this rapidly increasing specificity represents the unfolding of a complex, genetically inherited face processing mechanism, or some special kind of face-specific learning. Infants grow up in a world where satisfaction of their needs is far more likely to follow eye contact with a human face than to follow a glimpse of averted gaze, a chimpanzee's face, or a multi-colored bouncing ball. Therefore, domain-general processes of associative learning are sufficient to explain why, in the first year of life, a simple preference for inverted triangles of blobs (Figure 3.1) becomes a highly robust and selective preference for fellow humans "looking at me" (Heyes, 2016b).

Associative learning and, more specifically, reinforcement learning also explain how infants, building on an inborn face preference, develop gaze-cuing: a tendency to direct attention to the object, or area of space, to which another agent is attending (Moore and Corkum, 1994). Gaze-cuing, as opposed to motion-cuing, first appears at two to four months of age (Hood, Willen, and Driver, 1998; Farroni, Mansfield, Lai, and Johnson, 2003) and immediately expands the infants' opportunities to learn from and through others. Now they are looking not just at faces but at the objects and events to which the eyes or whole faces are directed. The flow of information that infants receive about the world is guided by adults' knowledge of what is important and interesting. Later, at six to twelve months, gaze-cuing becomes more selective and more active. Infants may become more inclined to follow gaze when the gaze shift is preceded by direct eye contact (Senju, and Csibra, 2008). They begin to look back and forth between an adult and an object, checking whether they have got the right spatial target, and to use pointing to bring the focus of an adult's attention into alignment with their own (Carpenter and Call, 2013). These "joint attention" behaviors may or may not reflect a growing understanding that other agents have mental states, or an increasing motivation to "share" those states (Tomasello, 2014; see Chapter 7). What is clear is that joint attention behaviors further increase the precision of social learning, make infants more teachable, and do not require Big Special cognitive processes for their development. As long as gaze shifts after eye contact (and fixations after checking) are more likely to yield an encounter with an interesting object, and as long as pointing tends to make adults do what infants want, reinforcement learning can build joint attention on the foundation of a simple, genetically inherited face preference (Moore and Corkum, 1994; Paulus, Hunnius, Vissers, and Bekkering, 2011; Triesch, Teuscher, Deák, and Carlson, 2006).[3]

Voices

Turning from vision to audition, there is evidence that humans start out with an input bias favoring voices. Like the inborn face preference, the voice preference is initially non-specific. At birth, infants will suck harder to hear speech sounds than to hear synthetic sounds with a similar pitch contour and spectral properties (Vouloumanos and Werker, 2007). However, human newborns also prefer rhesus monkey vocalizations to synthetic sounds, and suck equally hard to hear human speech and monkey vocalizations (Vouloumanos, Hauser, Werker, and Martin, 2010). Thus, at birth, the voice preference is not specific to human vocalizations—it encompasses at least some other primate calls—and it is only when babies are three months old that they are willing to work harder to hear human speech than monkey vocalizations.

In principle, this sharpening of the voice preference in the first few months of life could be due to the unfolding of a complex genetic program that specifies many acoustic and phonetic properties of human speech. However, a more likely explanation is that domain-general learning—the kind that produces habituation and sensitization to all kinds of stimuli, across species (Rankin et al., 2009)—makes infants enjoy speech more as they hear more of it. This is likely because, as ample evidence shows, newborns prefer speech sounds they heard while in the womb: for example, their mother's voice compared with other female voices (DeCasper and Fifer, 1980); the rhythms of their native language over those of a foreign language (Moon, Cooper, and Fifer, 1993); and stories that were read to them prenatally (DeCasper and Spence, 1986). No "genetic program" could include information specific enough to produce these preferences. Rather, these input biases show that domain-general processes of learning begin to encode speech input as early as the third trimester of pregnancy; they are, therefore, fully capable of honing a simple, inborn preference for primate vocalizations into a highly specific input bias favoring speech in the infant's native language (Reeb-Sutherland et al., 2011).

Domain-general learning may also play a crucial, and rather charming, role in the development of infant-directed speech, or "motherese"—the tendency of adults to address infants using speech that has a higher pitch, a broader pitch range, a stronger rhythm, and a slower tempo than the speech typically used to address fellow adults (Fernald, 1991; Trainor, Austin, and Desjardins, 2000). It was once thought that the production of infant-directed speech is a highly specific genetic adaptation favored by natural selection because it facilitates language learning by emphasizing the lexical and grammatical structure of utterances. However, it has subsequently been discovered that, when infant-directed speech is compared with emotional adult-directed speech, rather than emotionally neutral adult-directed speech, infant-directed speech is distinctive only in being of higher pitch (Trainor, Austin, and Desjardins, 2000).[4] This suggests that adults do not use a whole bag of vocal tricks when addressing infants. We use the same pitch range, pitch contours, rhythm, and tempo used when we are feeling emotional while addressing an adult. It is just that we are more likely to feel emotional when addressing an infant. The one trick tailored especially for infants is the higher pitch of infant-directed speech. Our tendency to use a higher pitch when talking to infants may or may not be a genetic adaptation. Here is the charming part: Smith and Trainor (2008) found that mothers could be "shaped" by their four month old infants to increase the pitch of their vocalizations. When mothers were consistently rewarded for higher pitch by happier behavior from the infant, reinforcement learning increased the average pitch of the mothers' vocalizations.

Babies from a few days to several months of age look longer at faces producing infant-directed speech than at faces producing adult-directed speech (Cooper and Aslin, 1990). The discovery that infant-directed speech is distinctive only in having a higher pitch suggests that, like the basic voice preference, this genetically inherited input bias is non-specific. Rather than being tuned to a highly distinctive set of vocal characteristics that evolved for language training or teaching more generally (Csibra and Gergely, 2006), infants simply like high-pitched speech. And this simple, inborn preference may be evolutionarily ancient and widespread in the animal kingdom. Trainor and Desjardins (2002) found that higher pitch impedes rather than facilitates vowel discrimination, which is not good for the language theory. Interestingly, there is evidence that infant-directed speech is effective in controlling the behavior of dogs, horses, cats, and other nonhuman animals (Snowdon, 2004).

So, research with newborns and other young infants suggests that we genetically inherit attentional biases in favor of face-like and speech-like stimuli, which are swiftly honed by domain-general mechanisms of learning. These mechanisms detect that, in the vast majority of human social environments, the stimuli that are maximally predictive of reward—milk, warmth, cuddles, a glimpse of an interesting object—are not merely face-like or speech-like; they are human faces "looking at me" and producing speech with highly familiar phonetic properties. Consequently, the attention of infants becomes locked-on to exactly those agents who are able and ready to supply the information they require, both to meet immediate needs, and to become mature, adult members of their society.

COGNITION

Dual-process models have provided a framework for research on cognition ever since psychology became an empirical science (James, 1890), and they continue to inspire some of the most rigorous, cumulative work in the field (Evans and Stanovich, 2013; Kahneman, 2003). These models vary in detail, but they are united in suggesting that cognition, and especially human cognition, is controlled by two systems, or sets of processes, that interact with one another. The operation of System 1 is typically characterized as fast, automatic, parallel, and based on information derived from genetic inheritance (for example, producing inborn attentional biases, like those discussed above) and from domain-general processes of learning (for example, motor skills, such as those involved in driving a car, that have become habitual). The operation of System 2 is slow, effortful, serial, and based on information both from System 1 and generated by its own activity. System 2 acts as a more or less successful "supervisor" or "executive" with respect to System 1 (Norman and Shallice, 1986); it schedules, harnesses, and augments the activities of System 1.[5] Dual-process theorists typically assume, without comment or elaboration, that System 1 is phylogenetically ancient, whereas System 2 has undergone major expansion in the hominin line.

System 1—Associative Learning

Associative learning can be very broadly defined as learning in which an excitatory or inhibitory link is formed between representations of events. The events can be changes in the animate or inanimate environment, perceived via any sensory modality ("stimuli"), or actions performed by the learner ("responses"). Associations can be formed between stimulus representations (stimulus-stimulus), between stimuli and actions that occur in the presence of those stimuli (stimulus-response), and between actions and their outcomes (response-reinforcer). The event representations are "lean," merely sensorimotor images of the events themselves or, in more neurological terms, reactivations of the circuits involved in perceiving or enacting the events.

Research on associative learning originated some 300 years ago in the work of the British Empiricist philosophers, John Locke and David Hume, who argued that human knowledge is based on sensory experience. Associative learning became the focus of an experimental research program with Ivan Pavlov's work on dogs in the early part of the twentieth century, and this program has been progressing ever since. Experiments on associative learning often use classical (or "Pavlovian") and instrumental (or "operant") conditioning procedures with human and nonhuman animals. In a classical conditioning procedure, the participant is exposed to a relationship between two stimuli—for example, the sound of a bell and the presentation of food, or, in the human case, evidence that certain allergic reactions have occurred after the consumption of particular foods. In an instrumental conditioning procedure, the participant experiences a relationship between an action and an outcome—for example, a rat might be given food each time it presses a lever, or a human might be given points in a computer game each time she clicks a particular object. Increasingly, studies using classical and instrumental conditioning procedures not only record changes in participants' behavior, but also use functional magnetic resonance imaging to record brain activity, as well as computational modeling of both the behavioral and the brain imaging data to test hypotheses about associative learning (Boakes, 1984; Pearce, 2013).

Evidence of associative learning has been found in every vertebrate and invertebrate group where it has been sought, and in a wide range of functional contexts, from foraging to predator avoidance, mate choice, and navigation (Heyes, 2012b; MacPhail, 1982; Shettleworth, 2010). Cross-species comparisons suggest that genetic evolution has made some major qualitative changes to associative learning in the course of its multi-million year history (Dickinson, 2012). It is likely that associations were originally formed on the basis of contiguity alone; an association was formed between the representations of any pair of events that occurred together in time. Later, in some lineages, the process became dependent on prediction error. For an association to be formed, a pair of events still had to occur close together in time, but, in addition, one event had to be predictive of the other (Rescorla and Wagner, 1972). For example, to establish an excitatory link between representations of A and B, the probability of B occurring with or shortly after A must be higher than the probability of B occurring in the absence of A. Finally, but still many millions of years ago, the process evolved such that prediction error modulated not only the rate at which associations were formed, but the "associability" of events. Representations of events that proved to be good predictors in the past were more likely to enter into associations with other events in the future (Mackintosh, 1975). With each of these changes, associative learning became better able to fulfil its biological function of tracking causal relationships between events (Dickinson, 2012).

There is no evidence to suggest that associative learning has undergone any major, qualitative changes in the recent past, and certainly not in the hominin line. However, it is plausible that, compared with other apes, humans genetically inherit an enhanced capacity for associative learning. We may be genetically prepared to forge associations faster, learn more of them in parallel, and / or more readily attach associations to specific contexts (Holland, 1992)—for example, learning that waving at a taxi makes it stop, but only if the taxi's light is on. These possibilities are very difficult to assess empirically because when one species learns more or faster than another, it is often due to genetically inherited differences in their perceptual, motor, attentional, or motivational systems, rather than to more efficient learning mechanisms. Furthermore, in adult humans, efficient associative learning could be due to the use of stimulus processing strategies, such as "chunking"—grouping stimuli into larger units—devised by the learner's own System 2, or learned via System 2 from other agents (Heyes, 2016c; Shea et al., 2014). However, indirect evidence that we humans are genetically prepared to learn more associations than our ancestors comes from a study showing that baboons and pigeons use the same processes to encode picture-response associations; but, compared with pigeons in the same learning environment, baboons can learn four times as many associations—around 4,000 in total (Fagot and Cook, 2006). This result suggests that genetic evolution has enhanced associative learning capacity in the course of primate evolution, thereby making it plausible that further expansion occurred in the hominin line.

The hypothesis that humans have a genetically enhanced capacity for associative learning is certainly consistent with the evidence that this kind of learning underpins not just "spit and twitches" (Rescorla, 1988) but many complex aspects of human behavior and decision-making. For example, associative learning makes a substantial independent contribution to performance on standardized tests of intelligence (IQ; Kaufman, DeYoung, Gray, Brown, and Mackintosh, 2009); mediates the development of our sense of agency, in other words, the phenomenal experience of producing events through one's own intentional action (Moore, Dickinson, and Fletcher, 2011); and enables us to use geometric cues in navigation (Prados, 2011). Furthermore, associative learning plays a fundamental role in learning about relationships between actions and their outcomes (Jensen et al., 2007; Shanks, 2010); in learning higher-order relationships between outcomes (Wunderlich, Symmonds, Bossaerts, and Dolan, 2011); in updating percepts (Li, Howard, Parrish, and Gottfried, 2008); and in tracking the value of social cues, such as advice, as well as asocial cues, in decision-making (Behrens et al., 2008; Garvert, Moutoussis, Kurth-Nelson, Behrens, and Dolan, 2015; Heyes, 2012b).

Newborn human infants show classical and instrumental conditioning (Blass, Ganchrow, and Steiner, 1984; Siqueland and Lipsitt, 1966; see also Emberson, Richards, and Aslin, 2015). These vintage studies confirm that the mechanisms of associative learning are genetically inherited, while more recent work with infants underlines the cascading effects of early associative learning and, therefore, its power to influence learning throughout development. In the first year of life, associations are learned "exuberantly" and remembered for long periods (Giles and Rovee-Collier, 2011; Rovee-Collier and Giles, 2010); earlier associative learning accelerates later associative learning (Bock, Poeggel, Gruss, Wingenfeld, and Braun, 2014; Rovee-Collier, Mitchell, and Hsu-Yang, 2013); and individual differences in rate of associative learning measured at one month of age predict social, imitative, and discriminative skills two, eight, and eleven months later (Reeb-Sutherland et al., 2011; Reeb-Sutherland, Levitt, and Fox 2012). One of the consequences of this cascade is rapid word learning—associations of sounds with referents—in the second year of life (Smith, Suanda, and Yu, 2014).[6]

Given the many roles played by associative learning in infancy and adulthood, an expanded capacity for this kind of learning is very unlikely to have been favored by genetic evolution for just one "reason." However, a notable consequence of being able to learn many associations in parallel is more precise categorization. When presented with a picture they have never seen before, pigeons that have learned many associations between pictures of trees and a certain response (for example, pecking a particular "Object A") are more likely to categorize the new picture correctly (to peck Object A if the picture contains a tree, and Object B if it does not) than pigeons who have learned fewer tree-response associations (Herrnstein, Loveland, and Cable, 1976). This kind of generalization is based on observable features of stimuli (branches, twigs, leaves) and enabled by simple mechanisms—pigeons do not acquire the concept of a tree through this kind of training (Chater and Heyes, 1994; Pearce, 2013)—but precise categorization provides a solid foundation for the acquisition of language-borne concepts. For example, a child who can categorize precisely the arm movements that do and do not result in contact between a hand and an object is well placed to master the concept of "reaching" and, from there, the concept of "desire" (Heyes and Frith, 2014; see Chapter 7).

System 2—Executive Function

Executive function has three core components: inhibitory control, working memory, and cognitive flexibility (Diamond, 2013). Inhibitory control involves the modulation of attention, behavior, thoughts, and emotions to override internally or externally generated impulses arising from the operations of System 1. Working memory holds and actively processes information about objects and events that are not currently present. (But beware, some researchers define working memory in a way that makes it virtually synonymous with the whole of executive function [see Duncan, Schramm, Thompson, and Dumontheil, 2012].) Cognitive flexibility is the capacity to switch between behavioral goals or among perspectives on a problem. It draws on inhibitory control because the preceding goal or perspective has to be inhibited, and on working memory because the new goal or perspective must be loaded into this mental workspace. The activities that are described in everyday life as "reasoning," "problem solving," and "planning" mature later and are constructed from the core components of executive function—inhibitory control, working memory, and cognitive flexibility (Collins and Koechlin, 2012; Lunt et al., 2012).

The most commonly cited evidence that genetic evolution has enhanced executive function in the hominin line comes from neuroanatomical studies showing that the prefrontal cortex of the brain, which is focally involved in executive function, is disproportionately larger in adult humans than in adult chimpanzees (Rilling, 2014; Passingham, 2008; Passingham and Smaers, 2014). Furthermore—and this is important given that later-evolving cognitive functions typically depend on many brain areas (Alvarez and Emory, 2006; Anderson and Finlay, 2014)—there is evidence that in humans, compared with other apes, the prefrontal cortex is more extensively connected with phylogenetically older, more System 1, parts of the brain, such as anterior cingulate cortex and the anterior insular (Peterson and Posner, 2012; Zilles, 2005).[7]

The power of human executive function is likely to be due in significant measure to learning, including social learning, but there is evidence that genetic inheritance also makes a substantial contribution. Studies of human individual differences show that performance on executive function tasks, and especially tests of working memory, correlate with performance on standardized tests of general intelligence, "*g*," and individual differences in *g* are genetically heritable (Bouchard, 2014; Duncan, Schramm, Thompson, and Dumontheil, 2012). In addition, comparative phylogenetic analyses indicate that *g* has undergone significant change in the course of primate evolution (Fernandes, Woodley, and te Nijenhuis, 2014; Reader and Laland, 2002).

It is likely that genetic evolution has expanded and refined, rather than remodeled, executive function in the hominin line because each of the three core components is present in extant nonhuman animals. A large-scale experimental study found varying degrees of self-control, a type of inhibitory control, in thirty-six species, including nonhuman apes, Old World and New World monkeys, carnivores, rodents, and birds (MacLean et al., 2014). For example, after seeing food hidden at a novel location, the tested animals could resist the impulse to search for the food at another location where they had seen it hidden repeatedly in the past. Similarly, there is evidence that rats and other rodents have working memory (Cook, Brown, and Riley, 1985; Matzel and Kolata, 2010), and rhesus monkeys are known to display a human-like performance profile in task switching experiments, suggesting that they and other nonhuman animals are cognitively flexible (Caselli and Chelazzi, 2011).

Each of the three core components of executive function is slow to develop in childhood and can be improved by training throughout the lifespan (Diamond, 2013). In Western samples, the slowest component, inhibitory control, is still maturing in adolescence (Luna, 2009). This combination, slowness and trainability, indicates that, in addition to genetic inheritance, learning and cultural inheritance play major roles in the development of human executive function. Whenever a child is taught to slow down to improve accuracy, to look away from a source of temptation (inhibitory control), or to break into chunks a set of things-to-be-remembered (working memory), the child is culturally inheriting a strategy to improve executive function. Yet more important from the starter kit perspective, our genetically inherited potential to develop powerful executive function supports the cultural inheritance of cognitive gadgets. As executive function develops, it becomes possible to upload mills (cognitive mechanisms) from the sociocultural environment, including the mills that enable high fidelity–high bandwidth cultural inheritance, such as metacognitive social learning strategies (Chapter 5), mindreading (Chapter 7), and language (Chapter 8).

CONCLUSION

In this chapter, I have surveyed some of the evidence that our genetic starter kit consists not of Big Special cognitive processes but of refined and expanded versions of the kits that are genetically inherited by chimpanzees and other animals. Compared with other extant primates, we are less aggressive to conspecifics and more strongly motivated to interact with them. These emotional and motivational features give developing humans more and better access to models and teachers, and make juveniles more malleable in interaction with others. We also have attentional biases, in favor of faces and voices, as well as biological motion, equipping us to begin extracting information from other agents as soon as we arrive in the world. These biases are simple and crude at birth, but, through the action of associative learning, they swiftly become so specific that they target our attention—and thereby our learning—on knowledgeable adults of our own cultural group, and on the objects and events to which those adults are attending. Finally, compared with other animals, we genetically inherit expanded capacities for associative learning and executive function. These cognitive processes allow us to process the information about the world flooding in from other agents, and to build new cognitive processes that, among other things, further enhance our ability to learn from others.

This picture, and the one I hope to draw in the book as a whole, is of mighty oaks growing from little acorns. The oaks are the Big Special cognitive processes found in mature adult humans. Each acorn is a genetically inherited starter kit consisting of temperamental factors, attentional biases, and the potential to develop especially powerful domain-general processes of learning, memory, and cognitive control. This picture suggests that High Church evolutionary psychology, and therefore cultural evolutionary theory, has underestimated the potential of the acorns discussed in this chapter, not primarily because it has overestimated the size of the oaks, but because it has failed to appreciate what domain-general cognitive processes can achieve when they are souped-up, tightly constrained by inborn attentional biases and prior learning, and immersed in a rich sociocultural environment.

This chapter, and the two that preceded it, have laid some foundations for cultural evolutionary psychology. They have located this approach relative to other answers to the question, "Why are humans such peculiar animals?" (Chapter 1); introduced a teleosemantic conception of the nature-nurture debate; explained, in broad terms, how selectionist cultural evolutionary theory can be applied to cognitive mechanisms (Chapter 2); and discussed some distinctively human psychological characteristics that we have good reason to believe are genetically inherited (Chapter 3). We will return to the big picture in the final chapter. In the meantime, we will focus on a series of "case studies." Chapters 5–8 each concern a distinctively human cognitive faculty that, I argue, has been shaped by cultural, rather than genetic, evolution. All of these faculties—selective social learning, imitation, mindreading, and language—are types of social learning and, more specifically, of cultural learning. Chapter 4 introduces the case studies by distinguishing cultural learning from other types of learning and explaining why the difference matters.

4

CULTURAL LEARNING

CULTURAL EVOLUTIONARY PSYCHOLOGY IS BOTH a framework for research and a hypothesis. As a framework, it recognizes that distinctively human cognitive mechanisms can be shaped by culturally inherited information, as well as by genetically inherited information and learning (Chapter 2). As a hypothesis, it proposes that cultural inheritance has played the dominant role in shaping all or most distinctively human cognitive mechanisms (Chapter 1). In an attempt to persuade you of the merits of the hypothesis that distinctively human cognitive mechanisms really are constructed by cultural evolution, Chapters 5–8 each look in detail at evidence relating to one particular type of cognition.

It is essential to scrutinize the evidence from cognitive science because that is where the case for cultural evolutionary psychology originates. I am *not* saying that distinctively human cognitive mechanisms *must* have been shaped by cultural evolution, nor that alternative, genetically based accounts are implausible on general evolutionary grounds. For example, I am not denying that genetic assimilation happens in human populations, querying evolutionary psychology's assumptions about the "Pleistocene past," nor suggesting that mathematical modeling, constrained by archaeological data, shows that insufficient time has elapsed for genetic evolution to produce cognitive instincts. Rather, I am arguing that, although the cognitive instincts hypothesis is plausible enough on general evolutionary grounds, the data from cognitive science indicate that it is incorrect—that distinctively human cognitive mechanisms are gadgets, not instincts, constructed in the course of human ontogeny through social interaction.

Cognitive mechanisms that we have reason to believe are distinctively human—present in most adult humans and absent (or merely nascent) in other species—include: (1) mechanisms that are specialized for dealing with the inanimate world, such as causal understanding; (2) faculties that are equally likely to process animate (social) and inanimate (asocial) events, such as episodic memory; and (3) various forms of cognition specialized for dealing with social stimuli, such as face processing, imitation, and mindreading. From this smorgasbord, I have chosen for close examination four types of what has come to be known as "cultural learning" (Tomasello, Kruger, and Ratner, 1993): selective social learning (Chapter 5), imitation (Chapter 6), mindreading (Chapter 7), and language (Chapter 8). For each of these faculties, evidence indicates not only that the adult form is distinctively human and specialized for processing social stimuli, but that it also enables the cultural inheritance of information.

Cultural learning is especially important for two reasons. First, both evolutionary psychologists and cultural evolutionary theorists—although divided on many issues, including the number and identity of cognitive instincts—are united in assuming that mechanisms of cultural learning are genetically inherited (Chapter 1). Therefore, in the interests of scientific progress, cultural evolutionary psychology warrants pursuit as a descendant of, and alternative to, evolutionary psychology and cultural evolutionary theory only if it can challenge this consensus. Second, from the perspective of cultural evolutionary theory, which I broadly share, mechanisms of cultural learning play a crucial role in making human lives so different from those of other animals. Like other distinctively human faculties, cultural learning meets challenges that arise in an individual's lifetime, enabling each of us to navigate the world of people (for example, face processing) and things (for example, causal understanding) with remarkable agility. However, unlike other faculties, cultural learning also underwrites a whole new inheritance system: cultural evolution. It enables each person and social group to benefit from the accumulated experience of innumerable other people, past and present, and thereby collectively to acquire knowledge and develop skills that far surpass those of other species.

So, in laying foundations for cultural evolutionary psychology, there are good reasons to focus on the cognitive science of cultural learning. However, before plunging into the case studies in Chapters 5–8, we need to clarify what is meant by "cultural learning" and how it relates to other kinds of learning. Cultural evolutionary theorists, and others interested in social learning, tend to have disciplinary roots in anthropology, mathematics, behavioral ecology, or behavioral economics, not in cognitive science. Consequently, they have sliced and labeled individual learning, social learning, and cultural learning with reference to behavioral effects and folk psychological concepts. This needs to change if insights from cognitive science are to be added to the richly interdisciplinary field of cultural evolutionary theory. Cognitive science is needed not only for the project outlined in this book, bringing cultural evolutionary theory to bear on the origins of distinctively human cognitive mechanisms. It is also needed for existing projects, in which the products of cumulative culture are seen as providing the selection pressure for the genetic evolution of distinctive human cognition. Even according to that hypothesis, cognitive processes play key roles—as the towering achievements of human gene-culture coevolution and the drivers of the cultural evolution of grist—and therefore demand characterization consistent with contemporary scientific knowledge about the mind. The job can't be left to folk psychology and the vestiges of behaviorism (Heyes, 2016a; in press, a).

This short chapter first outlines the way cultural evolutionists are currently thinking about cultural learning, and then presents a new, subtly different framework, which makes room for cognitive science to contribute to our understanding of cultural evolution. Along the way, it introduces and cuts through the byzantine terminology often used to talk about social and cultural learning.

SLICING AND NAMING TYPES OF LEARNING

At the beginning of his book, *The Secret of Our Success* (2015), Joseph Henrich, a prominent cultural evolutionist of the California school, explains how he uses the term "cultural learning" and how, in his view, cultural learning relates to other kinds of learning. The passage is worth quoting in full, and studying closely, because it is an unusually clear and explicit statement of how cultural evolutionists, and others with a specialist interest in social learning, typically think about cultural learning.

> Throughout this book, *social learning* refers to any time an individual's learning is influenced by others, and it includes many different kinds of psychological processes. *Individual learning* refers to situations in which individuals learn by observing or

> interacting directly with their environment, and can range from calculating the best time to hunt by observing when certain prey emerge, to engaging in trial-and-error learning with different digging tools. So, individual learning too captures many different psychological processes. Thus, the least sophisticated forms of social learning occur simply as a by-product of being around others, and engaging in individual learning. For example, if I hang around you, and you use rocks to crack open nuts, then I'm more likely to figure out on my own that rocks can be used to crack open nuts because I'll tend to be around rocks and nuts more frequently and can thus more easily make the relevant connections myself. *Cultural learning* refers to a sophisticated subclass of social learning abilities in which individuals seek to acquire information from others, often by making inferences about their preferences, goals, beliefs or strategies and / or by copying their actions. When discussing humans, I'll generally refer to *cultural learning,* but with non-humans and our ancient ancestors, I'll call it *social learning,* since we often aren't sure if their social learning includes any actual cultural learning. (Henrich, 2015: 12–13)

This passage implies that the relationships between individual learning, social learning, and cultural learning are as shown in Figure 4.1. One set of psychological processes enables agents to "learn by observing or interacting directly with their environment," labeled "individual learning." Another, partially overlapping set of psychological processes enables agents to learn in a way that is "influenced by others," labeled "social learning." In the overlapping area, learning is social in that it is influenced by others, but it occurs via the same processes as individual learning. Outside the overlapping area is a "sophisticated subclass of social learning abilities in which individuals seek to acquire information from others, often by making inferences about their preferences, goals, beliefs or strategies and / or by copying their actions." This subclass, which may or may not fill the area of social learning that does not overlap with individual learning, is called "cultural learning."

My first worry about the method of slicing and naming represented in Figure 4.1 is that it does not include the simple, superordinate category "learning." If one wants to characterize types of learning—such as "social learning" and "cultural learning"—it seems natural to start by saying what you mean by "learning." This worry may be trivial, nothing but tidy mindedness on my part, or consequential. For example, the lack of explicit acknowledgement that "learning" is the superordinate category could have made cultural evolutionists slow to recognize that psychological research on learning is highly relevant to their interests. More specifically, it may have led cultural evolutionists to overlook evidence that all social and cultural learning involves some of the same processes as individual (or "asocial") learning; evidence that social and cultural learning depend not on entirely different processes, but on extra or specialized mechanisms of learning.

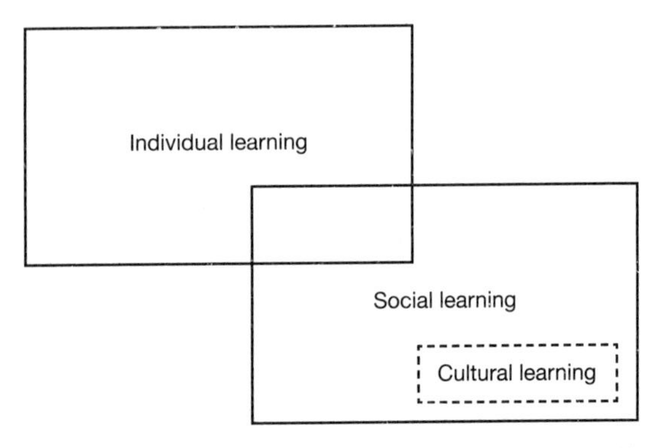

4.1 The received view of relations between individual learning, social learning, and cultural learning.

My second concern is that cultural evolutionists, and others with a special interest in social learning, tend to describe as "processes" phenomena that cognitive scientists would regard as "effects": things that need to be explained, rather than things that do the explaining. What does it mean to say that social learning encompasses "many different kinds of psychological processes"? A cognitive scientist would naturally assume that this phrase refers to a variety of subpersonal cognitive mechanisms that encode information for long-term storage—that is, learn—according to different rules. It is much more likely that Henrich was referring to "processes"—with labels like "stimulus enhancement," "local enhancement," "emulation," "mimicry," "response facilitation," "observational conditioning," and "imitation"—that social learning experts in behavioral biology distinguish according to *what* is learned, not *how* it is learned (Hoppitt and Laland, 2013). For example, a commonly used taxonomy defines stimulus enhancement as social learning in which the observer learns "to what (object or location) to orient behavior," and imitation as social learning in which the observer learns "some part of the form of a behavior" (Whiten and Ham, 1992). This taxonomy makes no mention of cognitive or neurological processes—processes between the agent's ears—that actually do the learning; it does not "enquire within" (Heyes, in press, a).

Finally, the most troublesome feature of Henrich's scheme in the present context: it does not offer a clear and coherent way of distinguishing cultural learning from other kinds of social learning. The quoted passage tells us that cultural learning is "sophisticated," not merely a "by-product of being around others," and it is active ("individuals seek to acquire information from others"). "Sophisticated" implies complex, but since we are not given any characterization of the processes underlying individual, social, or cultural learning—by any cultural evolutionist—it is impossible to guess on what dimension(s) cultural learning processes might be especially complex. The quoted passage says that cultural learning can involve "making inferences about . . . preferences, goals, beliefs or strategies" but does not explain what is meant by "inferences." In contemporary cognitive and behavioral science, the term "inference" is sometimes used in a strong sense, to refer to learning via a rule-governed process, using language-like mental representations, and sometimes in a very weak sense, as a synonym for all learning, including associative learning. The strong sense of "inference" would set a very high bar for cultural learning, whereas the weak sense would not distinguish cultural learning from any other kind of learning, social or otherwise.

There is more substance in the suggestion that non-cultural social learning is a "by-product" whereas cultural learning is active, but this distinction is liable to put many dumb animals in the culture club. For example, before consuming their first meal of solid food, weanling rats seek out a feeding adult member of their colony and eat whatever that adult is eating (Galef, 1971). This enables the weanlings to learn what is safe and nutritious to eat in their local environment. Thus, the weanlings engage in active social learning that depends crucially on what the adult rat is doing; it is not merely a by-product of proximity to adults, but it is unlikely that any cultural evolutionist would want the weanlings' behavior to be classified as an example of cultural learning. The whole point of circumscribing a subset of social learning as cultural learning—implicit in Henrich's book, and explicit when Tomasello and colleagues (1993) introduced the term "cultural learning"—is to isolate types of social learning that make the difference between the cumulative cultural inheritance found in humans, and the non-cumulative "culture" or "behavioral traditions" found in some other species. Cultural learning is supposed not only to support enduring regional variation in behavior, of the kind found in other animals (Whiten et al., 1999), but to allow tools and cultural practices to change in a directional way over generations, and even to improve—to get better at doing their jobs. A type of social learning found in both rats and humans may well contribute to cumulative culture in humans—it may play a role in carrying information from one generation to the next (Heyes, 1993)—but it is unlikely to be a "difference maker," a type of social learning that can help us explain why humans have cumulative culture and other animals do not. Therefore, it is not helpful to define cultural learning in a way that would make it widespread in the animal kingdom, or, as in the final sentence of the quoted passage, to bury the question of which types of social learning are distinctively human, under a convention in which learning from others is called social learning when it occurs in nonhuman animals, and cultural learning when it occurs in humans.

So, I have argued that the framework currently used to define cultural learning (Figure 4.1) has three shortcomings: (1) it does not acknowledge that cultural learning is, first and foremost, a form of learning; (2) it does not make contact with cognitive science because it focuses on observable behavior rather than the internal processes that generate behavior; and, most importantly, (3) it does not characterize cultural learning in a way that distinguishes it, conceptually or empirically, from other kinds of social learning.

AN ALTERNATIVE FRAMEWORK

Aiming to overcome these problems, I recommend the alternative framework shown in Figure 4.2. The superordinate category in this scheme is "learning"—encoding for long-term storage information acquired through experience.[1] When learning is assisted by contact with other agents, it is called "social learning." (This is the term most commonly used by researchers, but "copying," "imitation," and "public

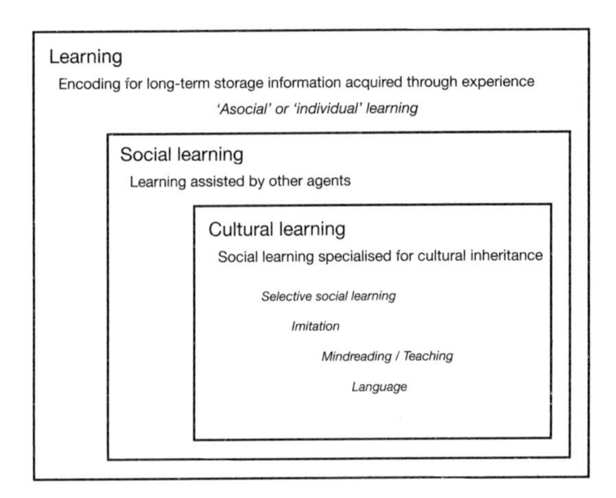

4.2 A framework for research on the relations between learning, social learning, and cultural learning enabling dialogue between cognitive science and cultural evolutionary theory.

information use" are popular alternatives.) When learning is not assisted by other agents, it is called "individual learning" (or "asocial learning," or "private information use"). "Cultural learning" is a subset of social learning involving cognitive processes that are specialized for cultural evolution—for example, they enhance the fidelity with which information is passed from one agent to another.

Notice that, within this framework, the distinctions between individual learning, social learning, and cultural learning hinge not on *what* is learned, but on *how* it is learned. In many parts of the world, turnstiles or "thigh-beaters" regulate access to train platforms and sports stadiums. When a person learns to use a turnstile—to get through without having their thighs beaten—they learn something that most people would say is not only social but cultural because it relates to an aspect of the built environment and varies across human populations. But learning to use a turnstile does not inevitably involve social learning or cultural learning. Turnstile learning is an example of both social learning and cultural learning when it depends on verbal instruction ("First put your ticket in this slot, then lean gently on the cross bar," and so forth). In this case, the learning is assisted by another agent, making it social learning, and involves a mechanism—language—that is specialized for cultural inheritance. Turnstile learning is an example of social learning, but not of cultural learning, when watching other people go through the turnstile draws the learner's attention to the ticket slot. In this case, the learning is assisted by others, making it an example of social learning, but via a domain-general attentional process, rather than a process specialized for cultural inheritance. Finally, turnstile learning is an example of individual learning, or simply "learning" (neither social nor cultural), when it occurs through solitary trial and error; when the learner, alone at the gate, tries various ways of getting through, and settles on the one that causes the least pain. In this case, the learning is not assisted in any way by another agent, let alone via a process specialized for cultural inheritance. Trial and error learning (also known as "instrumental learning" and "reinforcement learning") occurs in a wide range of animals, when they confront problems posed by social and inanimate features of the world.

My framework, outlined in Figure 4.2, is conventional in distinguishing social from individual learning according to whether other agents are involved, and in casting cultural learning as a special kind of social learning. However, two features of the framework make room for dialogue between cognitive science and research on cultural evolution. First, my scheme does not allude to "processes" in distinguishing individual from social learning, and therefore avoids the misleading impression that social learning is known to depend on different cognitive mechanisms from individual learning. Social learning has been carved into many slices by behavioral biologists, and these slices have been given some uninspiring labels (examples above), but the mere existence of these slices and labels does not imply that "stimulus enhancement," "observational conditioning," etc., are dependent on cognitive mechanisms distinct from those involved in individual learning. Indeed, as I shall argue in Chapter 5, there is evidence that most social learning—the kind that is not specialized for cultural evolution—depends on the same set of cognitive mechanisms as individual learning; it involves inputs from other agents, but those inputs are encoded for long-term storage via the same mechanisms as inputs from inanimate sources.

Second, although my scheme makes the conventional assumption that cultural learning involves processes specialized for cultural inheritance, it does not embody any assumptions about how or why these processes are specialized. Rather, it is a framework for investigating three questions that cultural evolutionists rarely tackle, or tackle effectively. First, the cognition question: How do the mechanisms of cultural learning differ from those of social learning at the cognitive level? Second, the contribution question: In what ways do the features that distinguish cultural learning from social learning contribute to cultural inheritance? (For example, do they make "improved" cultural variants more likely than "unimproved" variants to be passed on? Do they enhance the fidelity of their cultural inheritance?) And last, the specialization question: How have genetic evolution and cultural evolution contributed to the specialization of cultural learning?

Rather than appealing to "sophistication" or "inference"—implying that we already know what is distinctive about the mechanisms of cultural learning at the cognitive level (the cognitive question, above)—my framework defines cultural learning by ostension: by pointing at putative examples of cultural learning. The cultural learning box in Figure 4.2 lists the five categories of psychological phenomena (each containing behavioral effects and weakly specified cognitive processes) most commonly said by cultural evolutionists to be types of cultural learning: (1) selective social learning (also known as "learning biases," "transmission biases," and "social learning strategies"); (2) imitation (called "true imitation" when "imitation" is used as a synonym for "social learning"); (3) teaching (or "pedagogy"); (4) mindreading (also called "theory of mind," "mentalizing," "shared intentionality," and "social understanding"); and (5) language (so good they named it once). These five categories are a natural place to start asking the three questions of cognition, contribution, and specialization, posed above, about cultural learning. However, it is important to remember that the mere existence of a conventional category, such as "selective social learning," in no way guarantees that the category is coherent at the cognitive level and / or with respect to cultural evolution. For example, as I argue in Chapter 5, it could be that the behavioral effects in a single category, such as "selective social learning," are due to two different cognitive processes, X and Y, and only the effects of Y contribute to making human culture cumulative.

CONCLUSION

I recommend, and adopt in this book, a framework in which "social learning" is understood to be learning that is assisted in some way by contact with other agents (Heyes, 1994), and "cultural learning" is social learning specialized for cultural evolution. This approach is similar to, but less presumptive than, the framework currently used by cultural evolutionists. It makes clear that three important questions have barely been addressed by prior research on cultural learning, let alone resolved. The first is the question of *cognition*: How do the mechanisms of cultural learning differ from those of social learning at the cognitive level? The second is the question of *contribution*: In what ways do the features that distinguish cultural learning from social learning contribute to cultural inheritance? And the third is the question of *specialization*: How have genetic evolution and cultural evolution contributed to the specialization of cultural learning? The case studies in Chapters 5–8 address these three questions.

5

SELECTIVE SOCIAL LEARNING

"SOCIAL LEARNING" NAMES A RAGBAG OF BEhavioral effects, most of them found not only in humans, but in a wide range of animal species. At its most capacious, the bag contains all cases in which learning by one agent (the "observer") is influenced by contact with another agent (the "model" or "demonstrator") or its products. The influence can be on what, when, where, or how learning occurs. The learned information can be about the social or asocial world. The agents can be natural or synthetic—insects, birds, fish, rodents, cetaceans, primates, people, or robots—and the contact can involve anything from sniffing a slime trail to attending a lecture on calculus (Heyes and Galef, 1996; Hoppitt and Laland, 2013; Whiten and Ham, 1992).

Social learning is said to be "selective" or "biased," or the learner is said to be using a "social learning strategy," when the influence of other agents varies with the circumstances of the encounter ("when" selectivity), or with some feature of the available models ("who" selectivity). For example, there is evidence that learners may be more susceptible to social influence when the environment has recently changed (sometimes described as a *copy when uncertain* social learning strategy or bias) and more inclined to be influenced by older than younger models (*copy older individuals*) or by majorities than minorities within groups (*copy the majority*; Laland, 2004; Rendell et al., 2011).

Unlike imitation, mindreading, and language, selective social learning is not yet a major focus of psychological research, but it has played a prominent role in the California school of cultural evolution since that school's inception (Boyd and Richerson, 1985). Their modeling shows that selective social learning can influence how knowledge and skills spread within a population, and their commentaries imply that selective social learning helps to make cultural evolution adaptive, ensuring that "better" cultural variants are more likely to spread than "worse" cultural variants. However, as far as I can tell, the members of the California school have never spelled out how selective social learning does this (Clarke and Heyes, 2017). In common with developmental psychologists and behavioral biologists who study selective social learning, the California school assumes that the biases are cognitive instincts; social learning is made selective by genetically inherited cognitive processes.

Other putative types of cultural learning—imitation, mindreading, language—appear to be distinctively human, but selective social learning has been found in a wide range of species, including rats (Galef, Dudley, and Whiskin, 2008), sticklebacks (Webster and Laland, 2015), fruit flies (Battesti et al., 2015), and frog-eating bats (Jones, Ryan, Flores, and Page, 2013). The selective character of social learning helps to explain why social learning is adaptive and, therefore, why it is ubiquitous in the animal kingdom (Rogers, 1988). However, the evidence that social learning is always selective, in every species, presents a challenge for the view that selective social learning plays a key role in supporting human culture (Fogarty et al., 2012; Henrich, 2015; Laland and Rendell, 2013; Rendell et al., 2011). If selective social learning is a form of cultural learning, why is human culture so much richer than that of other animals? Why do rats, sticklebacks, fruit flies, and frog-eating bats not show human-like cultural diversity? Why do these species, and many others equipped with selective social learning, not show cumulative cultural change—the accretion of wisdom over generations, through social learning, to produce sophisticated technology, elaborate social practices, and vast libraries of knowledge about the world?

Surprisingly, cultural evolutionists rarely address this challenge. The one answer I have been able to find (Fogarty et al., 2012), which I will call the "memory hypothesis," implies several things. First, social learning is usually selective at the output stage. For example, observers acquire information from all available models and then decide on which source of information to base their behavior. Second, the output decision—for example, which model to copy—is made by psychological processes distinct from, and *on top of,* those that do the social learning. Third, the output decision processes, the candidate cognitive instincts of selective social learning, are guided by explicitly represented, domain-specific rules such as *copy the successful* or *copy older individuals.* Fourth, these rules or strategies can be more effectively deployed in humans than in any other animals because humans make more extensive use of memory. For example, when implementing *copy the successful,* humans keep track of an exceptionally long history of payoffs and use this record to detect change in payoffs, as well as to estimate the probability that such change is about to occur.

The memory hypothesis is not, and almost certainly was never meant to be, a satisfactory general account of why selective social learning has contributed to the emergence of cumulative culture in humans but not in other animals. For example, it is not clear that enhanced memory capacity would increase the efficiency of all or even most social learning strategies, and, if it did, this would implicate memory capacity, rather than selective social learning, as a difference-maker in relation to cumulative culture. However, the memory hypothesis provides an overview of how cultural evolutionists tend to think about selective social learning and, therefore, a point of departure for the alternative view to be presented in this chapter. Here, I wish to make several points. First, social learning is usually selective at the input stage rather than the output stage. For example, observers attend more closely to, and therefore learn more from, some models than from others. Second, the psychological processes that modulate input from other agents—broadly attentional processes—are distinct from those that do the social learning, but both sets of processes are domain-general; they are not specialized for processing animate rather than inanimate stimuli. Third, in humans, and only in humans, social learning is occasionally made selective by rules applied at the output stage, by "explicitly metacognitive social learning strategies"; in roughly the way that cultural evolutionists conceptualize all selective social learning. Fourth, explicitly metacognitive social learning strategies, such as *copy the boat builder with the largest fleet* and *copy digital natives* (people born since the advent of the Internet), promote cultural inheritance by enhancing the exclusivity, specificity, and accuracy of social learning. And fifth, explicitly metacognitive social learning strategies, like other metacognitive rules, are learned from other people. They are cognitive gadgets (Heyes, 2016a; 2016c).

The first section has a broad focus: it presents arguments and evidence suggesting that social learning, which is supposed to be made selective by social learning strategies, depends on the same core processes, the same processes of information encoding, as other (asocial) learning. Building on this analysis, the second section suggests that most of the selectivity found in social learning by nonhuman animals, children, and adults is due to modulation of the input to these basic learning mechanisms. It is neither uniquely human nor specialized for navigating the social world. The third section zeros-in on those rare cases of selective social learning, found only in humans, where there is reason to believe that the underlying mechanisms are guided by explicit, domain-specific rules. I argue that these mechanisms are metacognitive—cognitive processes that represent properties of other cognitive processes—and the rules are learned through social interaction. In the final section, I make some suggestions about how metacognitive social learning strategies promote cultural inheritance.

SOCIAL LEARNING

For historical reasons that need not detain us, most research on social learning has focused on nonhuman animals, and has been conducted by behavioral ecologists rather than psychologists. Consequently, there has been very little discussion of what a cognitive scientist would regard as psychological processes (see Chapter 4). However, there was, for many years, a widespread assumption that social learning, in humans and other animals, depends on different, genetically inherited processes from those mediating asocial learning (Klopfer, 1961; Templeton, Kamil, and Balda, 1999). This consensus has now evaporated, with many behavioral ecologists and psychologists believing that most social learning—perhaps all non-cultural social learning—depends on the same learning mechanisms as asocial learning, and that these are broadly associative processes that encode information for long-term storage by forging excitatory and inhibitory links between event representations (Dawson, Avarguès-Weber, Chittka, and Leadbeater, 2013; Frith and Frith,

2012; Leadbeater, 2015; Hoppitt and Laland, 2013). Four lines of evidence converge on this conclusion (Heyes, 1994; 2012c):

Social and asocial learning ability covary (Lefebvre and Giraldeau, 1996). Among birds (Reader and Laland, 2002) and among primates (Reader, Hager, and Laland, 2011), species that perform well in tests of social learning tend also to perform well in tests of asocial learning. This positive correlation is present even when statistical analyses control for body mass, brain volume, and phylogeny. There is also evidence that social and asocial learning covary across individuals within species (Boogert, Giraldeau, and Lefebvre, 2008; Bouchard, Goodyer, and Lefebvre, 2007).

Solitary animals are capable of social learning. In laboratory tests, animals such as octopuses (Fiorito and Scotto, 1992) and red-footed tortoises (Wilkinson, Kuenstner, Mueller, and Huber, 2010), which lead solitary lives in the wild, prove themselves to be adept at learning from social cues.

Social and asocial learning each have common varieties. Social and asocial learning each come in the same three basic flavors (learning about single stimuli, about relationships among stimuli, and about relationships between stimuli and responses, or actions and outcomes; see Heyes, 1994; 2012c), and each type of social and asocial learning has been found in a wide range of species including humans (Dawson et al., 2013; Leadbeater, 2015). For example, social learning about actions and outcomes, which was once thought to be uniquely human, has now been found in birds (Dorrance and Zentall, 2001; Saggerson, George, and Honey, 2005).

Social learning bears the footprints of associative learning. Relatively few studies have been designed explicitly to investigate the cognitive mechanisms mediating social learning, but research with this purpose has found signs that it is powered by associative mechanisms (Cook, Mineka, Wolkenstein, and Laitsch, 1985; Cook, Press,

Dickinson, and Heyes, 2010; Olsson and Phelps, 2007; Alem et al., 2016). For example, studies of human decision-making, combining mathematical modeling with functional brain imaging, have found that the same computations, based on the calculation of prediction error (the mismatch between expectation and incoming stimulation), are involved in processing information from social partners (social learning) and personal experiences of reward (asocial learning; Behrens et al., 2008; Garvert, Moutoussis, Kurth-Nelson, Behrens, and Dolan, 2015; Hill, Boorman, and Fried, 2016). Processing of social and asocial cues is sometimes carried out in different brain areas (Behrens, Woolrich, Walton, and Rushworth, 2007). However, evidence is emerging that, rather than being rigidly specialized for social and asocial learning, each brain area may be capable of processing social and asocial cues, switching back and forth according to which type of cue is currently more relevant for action (Cook, 2014; Nicolle et al., 2012; Rushworth, Mars, and Sallet, 2013).

Together, these four lines of evidence provide a strong indication that social and asocial learning are mediated by a common set of associative processes.

SELECTIVE LEARNING

Twenty years ago, it was radical to suggest that information is encoded for long-term storage by the same associative mechanisms in social and asocial learning. This idea is now embraced by many experts on social learning, and the debate about domain-specificity and adaptive specialization has moved elsewhere. Research is now focused on the possibility that there are cognitive mechanisms specialized by genetic evolution, not for the long-term storage of socially acquired information, but to make social learning selective. For example, processes that promote the use of information from certain types of models (who selectivity), or the use of information from other agents, rather than independently acquired information, under certain environmental conditions (when selectivity).

Strategic and Attentional Approaches

The mechanisms that make social learning selective have been conceptualized in two ways. The dominant conceptualization, which I will call the "strategic approach," implies that selection usually occurs at the output stage. If an observer is exposed to two models doing different things, she encodes both inputs– for example, she learns that both pushing and pulling a knob releases a catch—and then, when confronted with the knob, makes a decision about which action, pushing or pulling, to use herself. Thus, according to the strategic approach, it is performance rather than learning that is selective. Social learners take in all the information available but are choosy about how they use it to guide behavior. Such a selection mechanism is described as a decision made in accordance with a "rule" or "strategy," such as *copy the prestigious* or "prestige bias" (Laland, 2004; Henrich, 2015). Researchers who pursue the strategic approach avoid making explicit commitments about cognitive mechanisms (Hoppitt and Laland, 2013), but these formulae (*copy the prestigious*, etc.) imply that the mechanisms making social learning selective are high-level. To say that an agent "uses" a strategy implies a voluntary act rather than an automatic process. To characterize the strategy itself in words, such as *copy the prestigious,* implies that what is used is a rule encoded in a reportable form, and "copy" implies that the rule relates specifically to decisions about social learning, which is to say, that the strategies are domain-specific. Thus, although output selectivity could, in principle, be achieved by relatively low-level, automatic processes, those who pursue the strategic approach imply that, in the case of selective social learning, output selectivity depends on high-level processes.

The alternative "attentional approach" suggests that selection occurs at the point of information reception. If an observer is exposed to two models doing different things, she attends more closely to one than the other and, consequently, learns more about one action than the other. For example, she forges a stronger representation of the relationship between pushing and door opening than between pulling and door opening, and she is consequently more likely to push than to pull. A more prestigious model may well attract more attention than a less prestigious model, indicating a tendency on the part of the observer to pay closer attention to posh people than the hoi polloi—an empirical effect that can be described as "prestige bias." However, according to the attentional approach, the bias is due to modulation of learning by low-level or automatic attentional processes, not to the application of an explicit rule at the point when information is cashed out in behavior (Heyes, 2011; Leadbeater, 2015).

The strategic and attentional approaches are each consistent with the idea, outlined in the previous section, that the same mechanisms encode information for long-term storage irrespective of whether learning involves other agents; in other words, when learning is (descriptively) social and asocial. However, the strategic approach implies that mechanisms operate "on top" of these domain-general learning processes, and the higher processes are domain-specific, that is, dedicated to regulating when agents should use information acquired via a social channel, and on which models they should depend. Those who pursue the strategic approach also tend to assume that the domain-specific, higher-order selection processes are genetically inherited; that human prestige bias, for example, is a cognitive instinct. In contrast, the attentional approach assumes that social learning is made selective by mechanisms that are "upstream" of the domain-general learning processes—they come before learning—and highlights the possibility that these upstream attentional processes are also domain-general. They did not evolve specifically to modulate attention to social cues, and they select in the same way between and among social and asocial cues.

Research on domain-general attentional processes has shown that a stimulus is more likely to capture attention if it is perceptually salient (for example, if it is loud, bright, or occurs suddenly; Yantis and Jonides, 1990); or if the stimulus is relevant to an agent's emotional state, for example, anxiety and depression bias attention towards threatening stimuli (MacLeod, Mathews, and Tata, 1986), whereas positive mood biases attention towards attractive, rewarding stimuli (Tamir and Robinson, 2007). In these cases, attention to a particular stimulus "S" is influenced by the properties of S and the state of the agent at the time when attentional selection occurs. In other cases, attention varies with the prior consequences of attending to S. The effects of attending to S are more enduring when S has been followed by rich rather than meager financial rewards (Della Libera and Chelazzi, 2006), and stimuli that have proved to be good predictors in the past are given more attention than stimuli that have proved to be poor predictors (Le Pelley, 2010; Mackintosh, 1975).

For example, after completing a task in which stimuli vary on two dimensions (for example, color and line orientation) but only one dimension (for example, color alone) predicts the category to which a stimulus belongs, in a new localization task, where both dimensions are equally predictive, people show an automatic tendency to attend to the previously predictive stimulus dimension (color) rather than to the previously non-predictive stimulus dimension (line orientation; see Le Pelley, Vadillo, and Luque, 2013). Studies of this kind, in which adult humans and nonhuman animals are exposed to asocial stimuli, such as colors and lines, indicate that attention can be modulated by learning. Attention is captured not only by highly salient stimuli—cues that are loud, bright, or sudden—but by stimuli that are known on the basis of past experience to be good predictors of subsequent events. In these cases there is often a "learning sandwich": prior learning influences the distribution of attention across stimuli in a current scene, and this distribution influences what is learned by viewing the current scene or sequence.

Examples of Domain-General Selectivity

Most documented examples of selective social learning—in children, adults, and nonhuman animals—can be explained using the attentional approach. They are the effects one would expect if relatively low-level, domain-general processes—of the kind that have been investigated by cognitive scientists using arbitrary, asocial stimuli—also operate when agents are interacting with one another (Heyes, 2017a; Heyes and Pearce, 2015).

Here's a "who selectivity" example involving wild vervet monkeys (van de Waal, Renevey, Farve, and Bahary, 2010). In this study, vervets repeatedly observed a dominant female or male retrieving food from a box via one of two doors. When subsequently allowed access to the box themselves, observers of female models showed a stronger preference for the door used by the model than did observers of male models. Was this because the subordinates were equally likely to learn about the door-food relationship when observing male and female models, but more willing to copy the female (strategic approach), or because the subordinates attended more closely to the female model and, therefore, learned more about the door-food connection when observing a female (attentional approach)? Van de Waal and colleagues (2010) found support for the latter interpretation: using careful measures of attention, they found that observers were more likely to look at female than male models at the moment when they were opening the box. This result suggests that, at the level of psychological mechanisms, the female models were more effective than the male models because the females commanded more attention, and it is likely that this attentional bias was due to domain-general learning. In vervets, females are the philopatric sex—they remain in their natal group all their lives, whereas males migrate to another group when they are sexually mature. On average, therefore, females are more knowledgeable than males about the local environment. Consequently, most vervets have the opportunity to learn that female behavior has higher predictive validity than male behavior—watching females is more likely to tell you what will happen next—and thereby to learn an attentional bias in favor of females.

Up to the age of about five years, the "who selectivity" biases of human children are also explicable by the domain-general attentional approach (Heyes, 2017a). For example, it has been known for some time that children show prestige bias; they are more likely to copy a model that adults would regard as being of higher, rather than lower, social status (Bandura, Ross, and Ross, 1963; Harvey and Rutherford, 1960)—for example, their head-teacher rather than an equally familiar person of the same age and gender (McGuigan, 2013). Furthermore, recent studies have shown that children prefer models they have observed receiving social approval over those they have observed receiving disapproval or no feedback (Fusaro and Harris, 2008). Both of these "who" biases could be due to modulation of attention by domain-general processes of learning. Assuming that signs of social approval (looking, smiling, verbal expressions of approval) are rewarding for a child, when they are repeatedly directed toward a particular person, this pairing could—through an associative process known as "higher order conditioning"—make the observable features of the person more attractive (Rashotte, Griffin, and

Sisk, 1977). Consequently, the child will pay more attention to that person when they model an action, increasing the probability that the action will be learned.

Young children's "when selectivity" biases also fit the domain-general input modulation model. For example, Williamson, Meltzoff, and Markman (2008, Experiment 1) tested three-year-olds in a procedure with three stages: priming, observation, and test. In the priming stage, the children had an easy or a hard task, for example, to open a drawer that moved smoothly (easy), or a drawer that was jammed with putty (hard), to find a toy inside. In the observation stage, they were encouraged to watch an adult model performing the easy task using a distinctive method, for example, pressing a small button on the front of a second drawer before opening it smoothly and finding a toy. In the test stage, the children performed the task for a second time, but they were all given the easy version, for example, a third, unimpeded drawer in the same chest. The results showed that, on test, the children who had encountered the hard task at the beginning of the experiment were more likely than those who had encountered the easy task to copy the method used by the model, for example, to press the button on the front of the drawer.

Williamson and colleagues interpreted these results using the strategic approach—suggesting that the children copied selectively "in a rule-governed manner" (Williamson et al., 2008: 282)—but they fit the attentional approach exactly. Experiments on asocial learning in human and nonhuman animals have shown that the mechanisms of associative learning are driven by "prediction error"—the difference between what was expected to occur after an event (a stimulus or an action) and what actually occurred after the event (den Ouden, Friston, Daw, McIntosh, and Stephan, 2009; Schultz and Dickinson, 2000). Broadly speaking, the greater the prediction error, the more the agent attends to and learns about the relationship between the event and its outcome (Rescorla and Wagner, 1972). Therefore, the children in the experiment by Williamson and colleagues (2008) behaved exactly as one would expect if their behavior had been based on associative learning: they attended more to the behavior of the model at the observation stage and, consequently, learned more when there was a large prediction error at the priming stage—for example, they expected pulling to result in the drawer opening smoothly, but, instead, it jammed—than when there was a small prediction error at the priming stage—for example, they expected pulling to result in the drawer opening smoothly, and, although the trajectory may not have been exactly as they had anticipated from past experience with drawers, it did indeed open smoothly.

Even in adult humans, there is evidence that the selectivity of social learning is often due to low-level domain-general processes. For example, Behrens and colleagues (2008) gave adults a computerized task in which they scored points by selecting a blue or a green box in each of a large number of successive trials. In each trial, one of the boxes, blue or green, appeared in a red frame, indicating that it had been recommended for selection by an advisor. So, the participants could use personal and / or social information to make their decisions: in any given trial, they could base their decision on their own recent experience of the points value of each box (personal information) and / or select the box recommended by the advisor (social information). In practice, participants used both sorts of information in a highly flexible and adaptive way. In phases of the experiment when personal information was a better predictor than social information of the current value of the blue and green options, participants used personal information more than social information; and when social information was a better predictor than personal information, the participants relied more heavily on social information. Thus, viewed as an experiment on the selectivity of social learning, the results indicated a remarkably subtle form of "when selectivity"; over time and trials, participants relied on social learning to the extent that social cues did, and personal cues did not, predict decision outcomes. The crucial feature of the study by Behrens and colleagues is that it provided clear evidence, through mathematical modeling of brain imaging data, that the predictive value of both the personal cues and the social cues were tracked by mechanisms of associative learning. The value of personal and social cues was calculated using prediction error—a footprint of associative learning—in two different but neighboring areas of the anterior cingulate cortex, and when making a decision the information encoded by these parallel streams was combined within ventromedial prefrontal cortex.

These examples from studies of both nonhuman animals and human children and adults show that social learning can be made selective in the same way that other, asocial learning is made selective: by low-level, domain-general mechanisms, many of which modulate attention. Therefore, the mere fact that social learning is selective does not constitute evidence that there are mechanisms, genetically inherited or otherwise, dedicated to making it selective.

SELECTIVE SOCIAL LEARNING

To find evidence of dedicated social learning strategies—candidate forms of cultural learning—we need to find a home in cognitive science for the concept of a social learning strategy, and to look more closely at research on selective social learning in adult humans.

Explicit Metacognition

I suggest that the crucial, culture-relevant difference between selective social learning in humans and other animals is that *some* human social learning is made selective by explicit metacognition (Shea et al.,

2014); by conscious, reportable, domain-specific rules that represent "who knows," that is, properties of the cognitive processes of the rule user and other agents. Potential examples of these rules or "metacognitive social learning strategies," are *copy the boat builder with the largest fleet* and *copy digital natives.* Metacognitive rules focus social learning on knowledgeable agents so precisely that they encourage high-fidelity copying of behavior. Because it is exclusive, specific, and accurate, this kind of copying promotes cultural evolution by enhancing "parent-offspring relations" (Godfrey-Smith, 2012), keeping useful changes small and encouraging their proliferation to many agents (Heyes, 2016a; 2016c).

This metacognition hypothesis is intuitively appealing. Most of us can remember an occasion when we deliberated in a conscious way about "who knows," and on that basis decided whose example to follow, whom we should ask for advice, or whether we know enough to strike out on our own. Going well beyond intuition, formal evidence that social learning can be made selective by metacognitive social learning strategies comes from studies in which adult humans show selectivity that cannot be explained solely by the operation of low-level, domain-general cognitive processes, and that is in accord with their explicitly stated beliefs.

For example, in foraging and perceptual tasks, people have been asked to make a preliminary decision, and an explicit judgement of their confidence in that decision, before being given the opportunity to use social information to make a final decision (Morgan, Rendell, Ehn, Hoppitt, and Laland, 2012). The participants' confidence judgments were accurate—they had lower confidence in wrong than right preliminary decisions—and, crucially, they were increasingly likely to use social information as their confidence declined, suggesting that they deliberately applied the rule *copy when uncertain.*

Similarly, in another foraging task, people made use of social information—advice about which of two options to choose—to the extent that they believed the advisor to be motivated to help rather than to mislead them (Diaconescu et al., 2014). These beliefs were explicitly stated, and the basic effect—covariation between the advisors' incentives and the participants' use of social information—disappeared when participants were told that the advisors did not know which option they were recommending. Therefore, these results indicate that the participants used an explicitly metacognitive strategy such as *copy when the model intends to help.*

Metacognitive Strategies are Learned

The metacognition hypothesis suggests that the strategic approach to selective social learning is right, but only for a tiny minority of cases. In the vast majority of cases across the animal kingdom, and in young children, the selectivity of social learning is due solely to low-level, domain-general psychological processes. But occasionally, and only in humans, deployment of socially learned information really is regulated in a top-down fashion by explicit rules—rules that can be described literally in a formula such as *copy digital natives.* Identifying these rules as explicitly metacognitive forges a link between research on social learning strategies in behavioral ecology and economics, and research on metacognition in cognitive science. Metacognition is an increasingly popular and progressive topic in cognitive science (Fleming, Dolan, and Frith, 2012). Therefore, in time, this connection will tell us a great deal about the computational and neural bases of social learning strategies and, therefore, of cultural evolution.

An immediate implication of the metacognition hypothesis is that social learning strategies—real *metacognitive* social learning strategies—are likely to be culturally rather than genetically inherited. They are cognitive gadgets, rather than cognitive instincts. This follows from evidence that other, explicitly metacognitive rules are typically learned through social interaction (Bahrami et al., 2012; Hurks, 2012; Mahmoodi et al., 2013) and, therefore, show marked cross-cultural variation (Güss and Wiley, 2007; Heine et al., 2001; Li, 2003; Mayer and Träuble, 2013). For example, Western children learn by instruction to use "semantic clustering" to retrieve the names of animals from memory (for example, to think of birds first and then mammals; see Heine et al., 2001), and adults learn through social interaction explicitly to metarepresent their confidence in ways that make two heads better than one (see Bahrami et al., 2012).

Given these findings from research on metacognition in general, one would expect metacognitive social learning strategies to be products of learning through social interaction as well, and to vary across cultures. Consistent with the first of these predictions, I have been unable to find evidence of metacognitive social learning strategies in children below the age of four or five years, that is, before children have had the opportunity to learn such strategies by instruction (Heyes, 2017a).

Consistent with the second prediction, there is evidence of cross-cultural variation in the metacognitive social learning strategies used by adults (Efferson et al., 2007; Eriksson, 2012; Henrich and Broesch, 2011; Mesoudi, Chang, Murray, and Jing Lu, 2015; Toelch, Bruce, Newson, Richerson, and Reader, 2014). For example, in contrast with Westerners, Fijians are less likely to seek advice from people with more formal education (Henrich and Broesch, 2011), and, compared to Britons, people from mainland China engage in more social learning, and their social learning is less dependent on uncertainty (Mesoudi et al., 2015).

Thus, rather being genetically inherited, or devised afresh by each user, metacognitive social learning strategies are learned from others.

I didn't work out for myself that it's a good idea to *copy digital natives* when struggling with my computer. I was told to do it by a friend with teenage children, and since then I have adopted the strategy blindly, that is to say, without comparing the quality of the advice offered by younger and older models.

Metacognitive Strategies Promote Cultural Evolution

Chapter 3 outlined the dual-process view of cognition. Here I have presented a dual-process view of selective social learning, suggesting that most social learning strategies or biases are "planetary," based on domain-general psychological processes, while a few, found only in humans, are "cook-like," based on explicit, domain-specific rules (Heyes, 2016c). Planetary motion conforms to rules, but planets do not understand these rules or implement them deliberately; the rules of planetary motion are in the minds of scientists, not in the minds of planets. Similarly, the behavior of nonhuman animals and young children can be described and predicted by formulae such as *copy the successful* or *copy older individuals,* but the strategies or rules are in the minds of scientific observers, not of the actors themselves. By contrast, when people use explicitly metacognitive social learning strategies, they are like cooks rather than planets. Cooks know the rules to which their behavior conforms, and the conformity of their behavior is due, in part, to their knowledge of the rules.

The cook metaphor is especially apt because many recipes are products of cultural evolution. They are adapted over generations to local circumstances—climatic conditions, the availability of ingredients, fuel, and cooking vessels—and socially learned by the members of each new generation through observation of experts and active tuition at the campfire, in the kitchen, and, in some parts of the world, by reading recipe books. And, I have argued, the same is true for explicitly metacognitive social learning "recipes." The good ones—the metacognitive social learning strategies that accurately encode "who knows" within a particular social-epistemic ecology—are products of cultural evolution, learned through informal social interaction, in school, and by reading. People don't genetically inherit propensities to *copy the medicinal plant expert when you've got a health problem, and the fishing expert when you've got a net problem,* or to avoid herding by *copying the majority only when payoffs are visible* (Wisdom et al., 2013). Strategies or biases of this kind—the only kind that is truly strategic and specialized for cultural evolution—are learned from others by example and instruction. They are cognitive gadgets, not cognitive instincts.

The dual-process view of selective social learning is very far from deflationary. It suggests that even planetary social learning strategies are much more flexible and efficient than was previously thought. If, as behavioral ecologists and economists have assumed, social learning strategies were fixed products of genetic evolution, they would make social learning selective, but only in a way that was efficient in ancestral environments. For example, if older individuals tended to provide more reliable information in the distant past, agents alive today would be inclined to *copy older individuals* even if, in a tech-savvy world, younger individuals tend to know more, at least on certain topics. In contrast, because they are rooted in domain-general processes of learning, planetary social learning strategies can make social learning selective in a way that is adjusted rapidly, within lifetimes, to track changes in the social and asocial environment. Thus, if younger individuals provide more reliable information in particular contexts, agents will learn to attend to and copy younger individuals more than older ones in those contexts.

Because it casts even planetary social learning strategies as extraordinarily supple and adaptive, the dual-process view makes it difficult to identify the advantages of cook-like, explicitly metacognitive social learning strategies, and thereby to spell out what it is about distinctively human selective social learning that promotes cultural evolution. Here is an attempt to meet that challenge.

Step 1—From metacognition to better social learning strategies. Explicitly metacognitive social learning strategies are able to focus social learning on knowledgeable agents with greater accuracy and precision because they are themselves products of cultural evolution. Acquired through learning in the context of social interaction, metacognitive social learning strategies distill the accumulated wisdom of many agents about who knows best in various task domains. Planetary social learning strategies can be updated on the basis of only one agent's experience: the user's experience. If an agent gets higher payoffs when she copies younger individuals than when she sticks to asocial learning (or copies older individuals), she will develop—by planetary means, in other words, through domain-general mechanisms—a bias to *copy younger individuals.* But this bias only has a modest chance of being adaptive because it is narrow. It is derived from a relatively small sample of younger and older individuals—the small number of individuals that the focal agent has tried copying. In contrast, when a middle-aged person learns, by explicit instruction or via the zeitgeist, to *copy digital natives,* she is acquiring a metacognitive social learning strategy based on the payoff experience of a large number of other people—including all those who have been educated about information technology by their children, and have let this be known to others. Thus, whereas genetically inherited social learning strategies would be broad but inflexible, and planetary social learning strategies are flexible but narrow, metacognitive social learning strategies are both broad and flexible.

Step 2—From better social learning strategies to higher fidelity. Because metacognitive social learning strategies identify "who knows" with greater accuracy and precision, they increase the likelihood that agents will gain more by copying with higher than lower fidelity. In this context, fidelity has at least three components. First, exclusivity: deriving information from one or a small number of models, rather than by combining information from a large number of other agents (for example, *copy the majority*). Second, specificity: copying at a fine- rather than a coarse-grain, or exactly when, where, how, and in what order small components of the action are performed. Third, accuracy: copying without introducing random error or changes based on asocial learning (Goodnow, 1955).[1] When high fidelity copying is at a premium, metacognitive social learning strategies promoting exclusivity are favored by cultural evolution (for example, *copy the boat builder with the largest fleet* will gain more currency than *copy the majority's boat design*; see Godfrey-Smith, 2012), and both individual agents and social groups can afford to invest in the development of tools and cognitive mechanisms that allow copying with high specificity and accuracy. The cognitive mechanisms include executive processes focusing attention on the details of a model's behavior (and encoding the serial order of its components) and sensorimotor processes enabling translation of what has been observed into matching action by the observer.

Step 3—From higher fidelity to cultural evolution. Godfrey-Smith (2012) has argued that populational models of cultural evolution seek to explain the *distribution* of cultural traits—for example, why some ideas or skills are more common than others in certain social groups; or, in the case of selectionist models, both the distribution and the *origin* of cultural traits—for example, how particular skills, such as building a canoe from seal skin, could possibly come into existence (see Chapter 2). The success of distribution explanations—of applying models from population genetics to cultural phenomena—depends on good "parent-offspring relations" in the cultural domain. As in gene-based evolution, each new instance or "token" of a cultural type must be a copy of one or a small number of existing tokens. It is not sufficient for the earlier-occurring and later-occurring tokens merely to be alike, or for the latter to be loosely inspired by the former (Godfrey-Smith, 2012; Shea, 2009). My bread making skill is the offspring of your bread making skill to the extent that I acquired my skill by copying your technique and resisted blending your technique with others I observed (or with my own bright ideas about bread making). Therefore, if metacognitive social learning strategies promote exclusivity and accuracy in social learning (see Step 2)—if they reduce the number of models contributing to each new token of a cultural trait, as well as the degree to which the model's influence is contaminated by asocial learning—metacognitive social learning strategies will enhance parent-offspring relations, thereby increasing the power of populational models to explain the distribution of cultural traits. Or, to make the same point more directly: metacognitive social learning strategies help to create the conditions in which the distribution of cultural variants can evolve geographically over time.

Origin explanations—of how improvements in a cultural variant could accumulate to produce something as impressive as a seal skin canoe—are not as dependent on strong parent-offspring relations as distribution explanations. However, they do require that cultural variants change in small steps, and that useful new variants proliferate through the population in a way that creates many "independent platforms for further tinkering" (Godfrey-Smith, 2012). In other words, an impressive achievement can be ascribed to cultural evolution, rather than to the insight and ingenuity of a succession of individual agents, to the extent that each improvement was made more likely by the presence of *many* agents, rather than *smart* agents, using its precursor (Amundson, 1989). Therefore, both the specificity and the accuracy of social learning are relevant to origin explanations.

Specificity—copying at a fine- rather than a coarse-grain—helps to keep innovations small. Accuracy—copying with a minimum of random error or changes based on asocial learning—helps to ensure that small innovations proliferate intact to many agents within the population. There is always tension in Darwinian evolutionary models between variant generation and faithful retention (Campbell, 1974), but, to the extent that metacognitive social learning strategies support detailed and accurate copying, they are likely to help cultural evolution, rather than the smart choices of a succession of individual agents, to produce complex theories, artifacts, and practices.[2]

CONCLUSION

In the first section, I surveyed evidence that information is encoded for long-term storage by the same cognitive processes when the learning is either social or asocial—that is, when learning is and is not influenced by contact with another agent. This view is no longer controversial. Instead, those who regard social learning as "special" are now primarily concerned with selective social learning. They suggest that, in humans and a wide range of other animals, there are domain-specific processes—called social learning "strategies" or "biases"—that work in a top-down fashion to regulate the use of socially learned information. Challenging this view, I argued in the second section that selective social learning in nonhuman animals, in young children, and often in adult humans, is due to low-level, domain-general mechanisms: to the same, attentional processes that make all learning selective. However, in the third section I suggested that the strategic approach is occasionally right. In humans, social learning is sometimes made selective by metacognitive rules, that is, by culturally inherited, reportable generalizations about who is likely to have the best information, and when one should rely on that information rather than on one's own epistemic resources. The final section looked more closely at how the use of explicitly metacognitive social learning strategies, rules of the kind used by a cook, may contribute to cultural evolution.

Selective social learning has been a focus of cultural evolutionary theory since the 1980s, but it barely appears on psychologists' radar. In contrast, the topic of the next chapter—imitation—has been hailed by psychologists and biologists for more than a century as a distinctively human cognitive mechanism that plays a crucial role in supporting cultural accumulation.

6

IMITATION

IMITATION IS THE LAMBORGHINI OF SOCIAL LEARNING. There are many vehicles carrying information from one agent to another—with uninspiring names like "stimulus enhancement" and "observational conditioning"—but imitation has long been regarded as the sports car in the social learning garage. It has long been assumed that imitation involves complex computations specialized by genetic evolution for high fidelity cultural inheritance, and that this cognitive instinct, present at birth, plays a crucial role in allowing humans to make and use tools (Washburn, 1908).[1]

Imitation is impressive because it solves the "correspondence problem." The correspondence problem is tricky in two ways—it is difficult to see the problem and difficult to solve it. Even some experts on social learning don't "get" the correspondence problem. To see it, think about what is happening in Figure 6.1. The little boy's behavior is similar in a very specific way to the behavior of the men in front of him. It is "topographically" similar. That is, certain parts of the boy's body, his arms and torso, are configured—they are spatially related

6.1 An example of imitation. Downloaded from: https://www.catholicgentleman.net/2015/01/im-just-like-daddy/.

to one another—in the same way as corresponding parts of the men's bodies. In each case, the actor's arms are clasped behind the actor's back. This feature—topographic resemblance—is what defines the boy's behavior as imitation rather than some other kind of social influence. Imitation occurs when observation of a model causes the observer to perform a topographically similar behavior.[2] But very often, as in Figure 6.1, the topographic resemblance is visible only from a third-party perspective—for example, to us as viewers of the photograph. The little boy cannot see the resemblance. As he imitates the men, dipping his head and putting his hands behind his back, all the boy can see is the ground and perhaps the heels of the men in front. To the boy, hands-behind-back looks like ground-and-heels. The boy can feel his own movement—a slight tension in his shoulder joints, the touch of one hand on the other—but he cannot feel the movements of the men. And he can't compare his own movement with that of the men by listening because hands-behind-back does not produce a distinctive sound that is detectable to the human ear. So, how is the boy able to produce an action that resembles the men's action from a third-party perspective when he, the boy, cannot sense—see, feel, or hear—the resemblance? Or, to put the correspondence problem more generally: How does the cognitive system produce topographically matching actions when those actions are "perceptually opaque" (Heyes and Ray, 2000), that is, when disparate sensory inputs are received during observation and execution of the imitated action?

Most theories of imitation either do not address the correspondence problem at all, or suggest that it is solved by a uniquely human, genetically inherited cognitive mechanism—a mechanism that is able, by some unspecified means, to work out what it would feel like to perform every action we see, and that knows about the topographic correspondence between observed and executed actions. The first section of this chapter says a little more about these theories. The second outlines an alternative solution to the correspondence problem, the associative sequence learning (ASL) model of imitation. The ASL model, which is supported by more evidence than any other theory of imitation, suggests that the correspondence problem is solved by a culturally inherited cognitive mechanism. This cognitive gadget does not know about the topographic relationships between observed and executed actions. All it knows is that certain actions have been seen and felt together. The final section, "That Can't Be Right!" discusses objections to the idea that imitation is a cognitive gadget and argues that, although some of these objections are ingenious, none are compelling.

In short, this chapter embraces the idea that imitation is a Lamborghini, made possible by an impressive piece of cognitive kit, but argues that the kit is assembled from simple components in the course of development, and the "designer" was cultural rather than genetic evolution.

BLACK BOXES DELIVERED BY THE GENES

Andrew Meltzoff has provided an elegantly succinct characterization of the correspondence problem: How can the cognitive system "connect the felt but unseen movements of the self with the seen but unfelt movements of the other?" (Meltzoff and Moore, 1997: 179). His theory of imitation has dominated research since the late 1970s. At that time, Meltzoff and Moore (1977) began to publish evidence that newborn human infants, sometimes only a few hours old, can imitate a range of facial expressions and hand movements. Facial expressions are perceptually opaque—they yield very different sensory inputs when observed and executed—and newborns have had very little opportunity to learn about them. Therefore, these startling new data suggested that the cognitive mechanism that solves the correspondence problem is genetically inherited. As soon as the new data were published, other researchers began to raise questions and to report failures to replicate (Anisfeld, 1979; 2005; Jacobson and Kagan, 1979; Jones, 2006; 2007; 2009; Koepke, Hamm, Legerstee, and Russell, 1983; Masters, 1979; McKenzie and Over, 1983; Meltzoff and Moore, 1979). However, Meltzoff and Moore, and some other research groups, reported success in replicating and extending the original findings (for a review, see Ray and Heyes, 2011), and the idea that imitation depends on an inborn mechanism, or cognitive instinct, became the orthodox view in developmental psychology and beyond.

But what is the nature of this inborn mechanism? *How* does it convert "the felt but unseen movements of the self with the seen but unfelt movements of the other"? An answer to this question has never been given. The inborn mechanism is a black box—a device characterized only by its inputs (observed actions) and outputs (topographically similar executed actions). Meltzoff and Moore's "active intermodal matching" model describes the device as "innate equipment," says that it detects "equivalences between observed and executed acts," and suggests that the inborn mechanism codes both observed and executed acts "supramodally" as "organ relations"—the configuration of body parts produced by a body movement (Meltzoff and Moore, 1997; Meltzoff, 2002; 2005). However, the active intermodal matching model does not propose computations that would allow organ relations to be derived from observed body movements or to be cashed out as executed actions. So, the active intermodal matching model says there is an inborn thing inside the imitator that solves the correspondence problem, but it doesn't tell us how the thing works.

In cognitive neuroscience, not long before Meltzoff and Moore (1997) published their fullest exposition of the active intermodal matching model, Giacoma Rizzolatti and his group in Parma discovered neurons in the premotor cortex of monkeys with some very interesting properties (di Pellegrino, Fadiga, Fogassi, Gallese, and Rizzolatti, 1992). Now known as "mirror neurons" (Gallese, Fadiga, Fogassi, and Rizzolatti, 1996), each of these cells fires not only when a monkey executes a particular action (for example, pinching) but also when the monkey passively observes a similar action performed by another agent. Subsequent research, using single neuron recording and brain imaging, has confirmed that mirror neurons are also present in the premotor and parietal cortices of adult human brains (Molenberghs, Cunnington, and Mattingley, 2012).

The Parma group assumes that mirror neurons are genetically inherited and argues that their adaptive function relates to "action understanding" rather than imitation (Rizzolatti and Craighero, 2004). However, other researchers have been quick to assume that mirror neurons provide the neurological basis for imitation, and to interpret Meltzoff and Moore's evidence of neonatal imitation as a sign that mirror neurons are present at birth (Lepage and Theoret, 2007). There is now a good deal of evidence that mirror neurons are involved in imitation. For example, meta-analyses of imaging data have shown that mirror neuron areas of the brain are typically active when people are imitating (Caspers, Zilles, Laird, and Eickhoff, 2010), and disruption of these areas using transcranial magnetic stimulation selectively disrupts imitative behavior (Catmur, Walsh, and Heyes, 2009; Mengotti, Ticini, Waszak, Schutz-Bosbach, and Rumiati, 2013). Given this involvement, it is tempting to regard mirror neurons as a solution to the correspondence problem and, therefore, as an explanation of imitation. But this would be a mistake. "Involvement" is not explanation. Unless we know *how* mirror neurons solve the correspondence problem, they are just another black box. The question "How do people imitate?" becomes the question "How do mirror neurons imitate?" The target shifts but the challenges remain the same: to identify the source of the information that allows people and / or mirror neurons to relate the "seen but unfelt" to the "felt but unseen," and to explain how this information is handled such that people and / or mirror neurons produce behavior similar to that of the model.

IMITATION BY ASSOCIATION

Associative Sequence Learning

Some years ago, my research group set out to meet these challenges: to find a genuine solution, rather than a black box solution, to the correspondence problem. Our studies have involved adults, infants, and nonhuman animals, and used both behavioral and neurophysiological methods. The results of our studies, and those of other research groups, indicate that the correspondence problem is solved by a multitude of "matching vertical associations" like the one shown in Figure 6.2 (Catmur, Press, and Heyes, 2016; Catmur et al., 2009; Heyes and Ray, 2000). A matching vertical association is a sensory representation of an action linked to a motor representation of the same action. For example, a visual image of what it looks like when a person bows his or her head connected to a "motor image" of bowing, where the motor image both represents the feeling of bowing one's head, and has the power to launch a bowing action. The link between the sensory and motor representations is bidirectional and excitatory: activation of either representation increases the probability that the other will be activated. When people imitate, the direction is from sensory to motor. The sensory representation is activated by observing another person's behavior; this activation is propagated to the motor representation via the associative link between the sensory and motor representations; and the resulting motor activation enables (but does not compel; see Heyes, 2011) the observer to perform the observed action.

Matching vertical associations are forged by learning, predominantly social learning; they are not inborn or genetically inherited. Consider the baby shown in Figure 6.2A. Due to genetic inheritance and early experience of both observing and performing actions, she

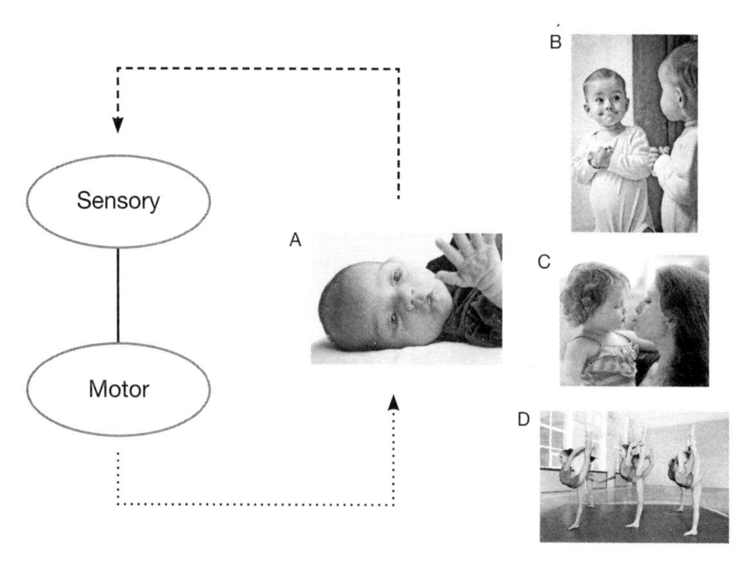

6.2 Matching vertical associations are acquired through sensorimotor learning. In the simplest case, self-observation (A), activation of a motor representation contributes to performance of an action (e.g., flattening the hand; dotted arrow), and observation of the performed action produces correlated activation of a corresponding visual representation (dashed arrow). Correlated activation strengthens the excitatory link between the sensory and motor representations, establishing a matching vertical association (solid vertical line). Opitical mirrors (B), being imitated by others (C), and synchronous activities (D) provide correlated sensorimotor experience for perceptually opaque actions, such as facial gestures and whole body movements.

is likely to have sensory and motor representations of a range of actions, and some of the sensory representations will be linked to motor representations. For example, a visual image of "nipple approaching" may be linked to a motor image of mouth opening. However, she will not develop matching vertical associations—excitatory links between sensory and motor representations of *the same* actions—unless or until she receives a very specific kind of experience. To develop matching vertical associations, she needs correlated sensorimotor experience: experience in which seeing and doing a particular action occur close together in time and in a predictive or "contingent" relationship. For example, a matching vertical association for grasping will develop if the infant is more likely to see grasping than to see any other type of action when she is grasping. For hand actions, this kind of experience can be obtained through self-observation (Figure 6.2A; White, Castle, and Held, 1964). When infants watch their own hands in motion, performance of each action is correlated with observation of the same action. Consequently, activation of a motor representation of grasping, involved in production of the movement, is correlated with activation of a sensory representation of grasping, provoked by viewing the movement, and, as research on associative learning has shown (Pearce, 2013), this correlated activation (contiguous and contingent) is necessary and sufficient to establish an excitatory link between the sensory and motor representations (James, 1890).

For perceptually opaque actions, such as facial expressions and whole body movements like hands-behind-back (Figure 6.1), self-observation does not provide the kind of sensorimotor experience necessary to build matching vertical associations. The location of our eyes relative to our faces and the rest of our bodies is such that, when we put our hands behind our backs, we cannot see ourselves as we would see or be seen by others. To develop matching vertical associations for perceptually opaque actions, we need not self-observation, but sociocultural experience—to interact with other people in culture-specific ways, or to interact with devices, like mirrors and video, that are products of cultural evolution (Figure 6.2B).

In cultures where adults engage in a lot of face-to-face contact with infants, imitation of infants by adults is an important early source of matching vertical associations for facial gestures (Figure 6.2C). In face-to-face interaction, Western mothers imitate their infants approximately once every minute (Pawlby, 1977; Uzgiris, Benson,

Kruper, and Vasek, 1989). Imitation of infants by adults enables matching vertical associations to be culturally inherited. For example, if a mother has a matching vertical association for frowning, it will be activated by the sight of her baby frowning and, in many cases, lead her to imitate the baby's frowns. When the baby is looking at his mother's face while he frowns, her imitative movements activate the baby's sensory representation of frowning, leading to correlated activation of the baby's sensory and motor representations of frowning and, thereby, to the establishment of a matching vertical association for frowning. Thus, the capacity to imitate frowning is passed on culturally, through social interaction, from mother to child.

Culture-specific forms of synchronous action are another important source of matching vertical associations for perceptually opaque actions, especially whole body movements (Figure 6.2D). When a group of people respond simultaneously to an external stimulus, such as a musical beat, in a prescribed way or with the same spontaneous gestures, each member of the group receives correlated experience of seeing and doing the actions in the sequence. Consequently, whenever groups of children or adults dance together, or perform synchronous drills, they are not only strengthening their social bonds (Tunçgenç and Cohen, 2016; Wiltermuth and Heath, 2009), but expanding and reinforcing their repertoire of matching vertical associations (Hove and Risen, 2009; Tarr, Launay, Cohen, and Dunbar, 2015). They are learning to imitate through culture-specific forms of social interaction.

Matching vertical associations solve the correspondence problem in two ways. They enable agents to imitate "familiar actions," such as frowning, that they could perform but not imitate before a matching vertical association was established. (This is sometimes called "mimicry" or "response facilitation.") They also allow agents to learn by observing how others perform "novel actions," such as

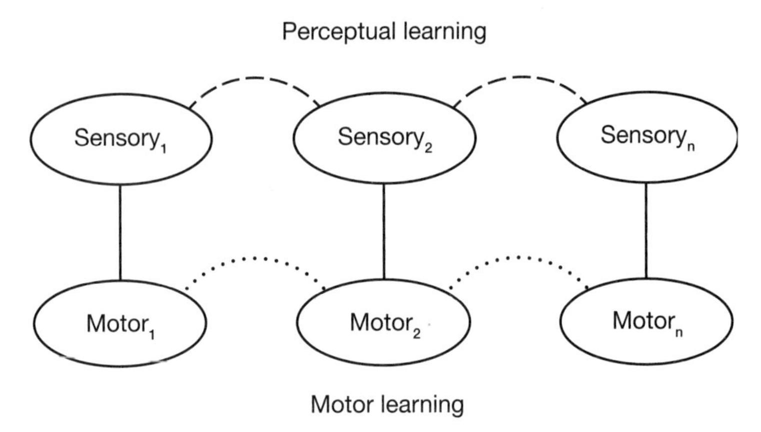

6.3 The ASL (cognitive gadget) model of imitation. Matching vertical associations (vertical lines) gear perceptual sequence learning (dashed lines) to motor sequence learning (dotted lines), creating a new, domain-specific cognitive mechanism that allows agents to copy the topography of observed body movements.

hands-behind-back, that were not previously part of the observer's behavioral stock. (This is sometimes called "true imitation," "imitation learning," or "observational learning.") According to the ASL model, matching vertical associations enable observational learning by connecting two processes—perceptual sequence learning and motor sequence learning—that normally operate independently (Figure 6.3). Perceptual sequence learning is a set of domain-general processes that encode the serial order of external stimuli. These stimuli can be in any sensory modality and originate from animate or inanimate objects. When a person watches the action of another agent, such as hands-behind-back, perceptual sequence learning encodes the order of the action components. (We are not good at describing body movement kinematics in words, because words are rarely involved in controlling them, but, in the hands-behind-back case, the components can be described roughly as: down—apart—swing—join.) Without matching vertical associations, this perceptual encoding of component order would allow the action to be recognized when seen again and provide a basis for inferences about the actor's intentions, but it would not allow the observer to perform the action him- or herself. However, when there are matching vertical associations for the components of the observed action, perceptual sequence learning drives motor sequence learning; that is, action observation drives processes that normally operate only when we are learning a new motor skill, such as riding a bike, through practice. The matching vertical associations act like teeth on the wheel of perceptual sequence learning, gearing it to motor sequence learning. Successive activation of sensory representations of sequence components leads, via the matching vertical associations, to successive activation of motor representations of the sequence components, and this allows the observer's motor system to encode the action sequence as if it were being performed by the observer.

In summary: the ASL theory contrasts with previous models of imitation in a number of ways. First, rather than a black box, the ASL model provides a mechanistic account of how the correspondence problem is solved for both familiar and novel actions. Second, it suggests that the mechanisms that solve the correspondence problem are relatively simple. Matching vertical associations do not calculate the degree of topographic resemblance between observed and executed actions. They just pair-up sensory and motor representations of action using excitatory links. Third, and most important for the theme of this book, the ASL model suggests that the mechanisms that enable us to imitate—matching vertical associations, the ones that solve the correspondence problem—are culturally rather than genetically inherited. A few matching vertical associations may be inborn (for example, smiling)—they may be "innate releasing mechanisms" (Lorenz and Tinbergen, 1938 / 1970)—but the vast majority are forged by sociocultural experience; that is, by interaction with other people in a context of culture-specific practices, and by exposure to artifacts such as mirrors. Even self-observation is culturally regulated. There is evidence that infants in the United States spend the majority of their waking moments watching their own hands in motion (White et al., 1964), but in cultures where babies are swaddled, the proportion is likely to be much lower.

Evidence

The evidence that inspired and supported the cognitive instinct theory of imitation recently suffered a mortal blow. It has long been suspected that a subtle experimental artifact has been giving the false impression that newborn humans can imitate (Anisfeld, 2005; Jones, 2009; Heyes and Ray, 2011). That suspicion has now been confirmed by a study conducted in Brisbane with unprecedented power and rigor (Oostenbroek et al., 2016; Heyes, 2016d). The Brisbane study tested more than a hundred infants, at four time points (one, three, six, and nine weeks of age), for imitation of eleven gestures, using the gold standard "cross-target" procedure introduced by Meltzoff and Moore (1977). In other words, the study asked, for each action tested, whether an infant who has just seen the action (for example, mouth opening) performed the same action with a higher frequency than other actions in the test set (for example, grasping). The answer was an unambiguous "no" for all actions apart from tongue protrusion.

In combination with earlier analyses (Anisfeld, 2005; Jones, 2009; Heyes and Ray, 2011), this result shows that newborns do not imitate, so why have we thought otherwise for forty years? Because most studies of neonatal imitation have tested tongue protrusion against one or two other target actions, and babies have a tendency to stick out their tongues just after they have seen tongue protrusion. This tendency is non-specific—babies also protrude their tongues when exposed to other arousing stimuli, such flashing lights and lively music (Jones, 2006)—but it has skewed the data from cross-target tests, giving the false impression that the capacity for imitation, and mirror neurons (Lepage and Theoret, 2007), are present at birth.

Reports of neonatal imitation didn't just inspire the cognitive instinct view of imitation; they were the only empirical evidence motivating and supporting that view. Consequently, the Brisbane study leaves no good reason to believe that the capacity for imitation is genetically inherited. In contrast, the ASL model, the cognitive gadget theory of imitation, is supported by evidence of two kinds (for reviews, see Catmur et al., 2009; Cook, Bird, Catmur, Press, and Heyes, 2014). The first comes from training studies showing that imitation—measured behaviorally and via "mirror responses" in the brain—can be changed profoundly by sensorimotor experience. The second indicates that imitation has "signature limits" (Butterfill and Apperly, 2013); it is constrained in the ways one would expect if imitation is controlled by matching vertical associations rather than by mechanisms that calculate the topographic similarity between observed and executed actions.

The *training studies*—involving adults, infants, and nonhuman animals—show that imitation can be enhanced, abolished, and reversed by novel sensorimotor experience. For example, adults usually do not imitate the actions of inanimate systems, such as robots, but after a brief period of training in which robotic movements are paired with topographically similar body movements performed by the observer, people imitate robots as much as they imitate other people (Press, Bird, Flach, and Heyes, 2005). And, in a complementary way, counter-imitation training, in which people perform one body movement while observing another, can eradicate (Heyes, Bird, Johnson, and Haggard, 2005), and even reverse (Catmur, Walsh, and Heyes, 2007; Catmur, Mars, Rushworth, and Heyes, 2011), a spontaneous or "automatic" tendency to imitate the trained actions.

For example, without training, passive observation of index finger movement activates muscles that move the index finger more than muscles that move the little finger. However, after training in which people are required to respond to index finger movements with little finger movements, and vice versa, this pattern is reversed. Observation of index finger movement activates little finger muscles more than index finger muscles, and vice versa. Automatic imitation is converted by sensorimotor learning into automatic counter-imitation.

Crucially, all of these studies controlled for the effects of perceptual learning, due to repeatedly seeing the trained actions, and of motor learning, due to repeatedly doing the trained actions. People in the control groups saw and did the actions just as often as people in the experimental groups, but they were exposed to a different relationship (matching or nonmatching) between what they saw and what they did. Therefore, the results show that, as the cognitive gadget view of imitation predicts, it is specifically sensorimotor learning, due to repeatedly seeing-and-doing, that has the power to enhance, abolish, and reverse imitation. The experiments have also included controls to check that the learning is associative, the same kind of learning that occurs in Pavlovian and instrumental conditioning experiments. Like other associative learning occurring in all vertebrate species, and in relation to inanimate as well as animate stimuli, the sensorimotor learning that changes imitation is context-dependent (Cook, Dickinson, and Heyes, 2012), and varies with contingency—the degree to which seeing the trained action predicts execution of that action, and vice versa (Cook et al., 2010; Cooper, Cook, Dickinson, and Heyes, 2013).

A recent study both illustrated the importance of contingency and confirmed that sensorimotor learning can induce imitative capacity in infants (De Klerk, Johnson, Heyes, and Southgate, 2015), in addition to changing imitation in adults. In this study, seven-month-olds who could not yet walk made stepping movements on a treadmill while intermittently observing those movements in real time on a video screen. Before and after this sensorimotor training, the babies passively observed the stepping movements of other infants while scalp electrodes recorded activity in the parts of their brains involved in generating leg movements. This mirror activity, indicative of a capacity to imitate, increased between the before and after tests only to the extent that the infants had experienced a contingency between seeing and doing stepping movements during training. Observing or performing a lot of stepping movements was not enough. The induction of mirror activity was related specifically to correlated sensorimotor experience, that is, to how often the infants had seen stepping while stepping rather than not stepping.

The *signature limits* research shows that, although the mechanisms underlying imitation are efficient and adaptive, they are not as flexible or as accurate as the cognitive instinct theory would lead us to expect. Indeed, they show just the kind of limitations that the cognitive gadget model predicts. For example, imitation learning is "effector-dependent"; it does not readily generalize across parts of the body. People who have observed a complex sequence of key-pressing movements can reproduce the sequence when their fingers are in the same keyboard positions as the fingers of the model, but they cannot imitate the sequence when their hands are crossed on the keyboard (right hand operates left keys, and vice versa), or when they are asked to use their thumbs rather than their fingers to press the keys (Bird and Heyes, 2005; Leighton and Heyes, 2010). Similarly, when people are asked to imitate a sequence of movements involving the selection of a pen and its placement in one of two cups using one of two grips, they show exactly the same pattern of errors as when they are instructed to perform the same movements by flashing geometric shapes; that is, when the task does not involve imitation (Leighton, Bird, and Heyes, 2010). Error patterns are indicative of underlying cognitive processes. Therefore, this finding indicates both that imitation is as error prone as a comparable "artificial" task and, as the ASL model says, that the sequence learning processes involved in imitation learning are domain-general—the same processes that encode the serial order of events in the inanimate world, and enable motor learning through practice rather than observation.

An experiment in which healthy human adults were asked to imitate their own facial expressions provides striking evidence that, when it comes to imitation, we're not as good as we think we are. First, the participants' faces were filmed while they were telling jokes. Then, several months later, they were brought back into the lab, shown stills from the movie, and allowed a series of attempts to reproduce the facial gesture they were making in each still. Precise, automated analysis of facial movements revealed that, unless the participants were able to see themselves while they were making these attempts—on a video screen functioning as a mirror—they did not show any improvement at all. Their first attempts to imitate were pretty bad, and their final attempts were no better (Cook, Johnson, and Heyes, 2013). The ASL model predicted this weak performance because in everyday life people rarely receive the sensorimotor experience necessary to establish matching vertical associations for subtle facial movements. Actors and impersonators spend a good deal of time looking in a mirror while speaking and laughing, but most of us use mirrors only for shaving or applying make-up, when our faces are relatively immobile.

In contrast, the cognitive instinct view predicted that the adults in this study would have no trouble improving their performance over trials. According to that theory, we should not need a mirror showing us our attempts because we have a powerful, genetically inherited black box inside our heads that can work out, for any action performed, what that action would look like from a third party perspective (Meltzoff and Moore, 1994; 1997).

THAT CAN'T BE RIGHT!

The previous section described a decisive study undermining the view that imitation is a cognitive instinct (Oostenbroek et al., 2016), surveyed the many studies that have tested the cognitive instinct and cognitive gadget accounts against one another, and found that the cognitive gadget theory provides a better fit with the data. Despite these findings, there is residual skepticism—a feeling among some specialists that, regardless of which way the data are pointing, the cognitive gadget theory just can't be right. To a degree, this skepticism is due to explanatory taste, a feeling that the cognitive gadget theory offers a prosaic, associative solution to a once-beautiful mystery and, therefore, that "the soul of imitation has been snatched away" (Meltzoff, 2005: 55). But, whatever the source of resistance, the cognitive gadget theory has been challenged in some interesting and important ways (Heyes, 2016e). Here are the most common objections.

Intervention versus Development

Objection: The course of a river can be changed by building a canal, but that doesn't mean the original course of the river was carved out by canal builders. Similarly, evidence that imitation can be changed by sensorimotor learning does not imply that the capacity to imitate normally develops through sensorimotor learning.

The logic of this objection is spotless. The statement "p implies q" means that if p is true, then q must also be true. If p is "imitation can be changed by sensorimotor learning," and q is "the capacity to imitate normally develops through sensorimotor learning," the truth of p definitely does not guarantee the truth of q. But that's just how it is in science. Scientific evidence supports inference to the best explanation, rather than deductive inference (Lipton, 2003). Empirical data do not guarantee the truth of any particular theory, but, if we are lucky, they favor one theory over another. The fact that sensorimotor learning can produce profound changes in imitative behavior favors the gadget theory over the instinct theory because it is exactly what the gadget theory predicts and difficult for the instinct theory to accommodate. If imitation is an instinct or module, one would expect stabilizing selection to have protected or buffered its development against environmental perturbations that could interfere with its adaptive function (Cosmides and Tooby, 1994; Pinker, 1997). One would not expect what the experiments show: that imitation can be diverted by relatively brief periods of sensorimotor learning, in which people are exposed to mismatched and arbitrary relations between observed and executed actions of kinds that are likely to have been encountered by our hominin ancestors.

It is also important to notice that the cognitive gadget theory of imitation does not rest on a single pillar. The instinct theory depended exclusively on evidence of neonatal imitation, but adult training experiments provide just one source of support for the gadget theory. For example, in addition to the work on signature limits, and the evidence that sensorimotor learning can induce mirror activity in infant brains (De Klerk et al., 2015), both outlined in the previous section, the gadget theory is supported by more naturalistic studies, many involving dancers and musicians (Calvo-Merino, Glaser, Grezes, Passingham, and Haggard, 2005; 2006; Cross, Hamilton, and Grafton, 2006; Ferrari, Rozzi, and Fogassi, 2005; Haslinger et al., 2005; Jackson, Meltzoff, and Decety, 2006; Keysers et al., 2003; Kohler et al., 2002; Margulis, Mlsna, Uppunda, Parrish, and Wong, 2009; Vogt et al., 2007). These show that imitative capacity varies with the extent to which participants have had the opportunity to acquire relevant matching vertical associations as they developed expertise in the real world (Cook et al., 2014).

Furthermore, and perhaps most important, the gadget theory is consistent with the timetable of imitation development revealed by observational studies of infants at home and in day nurseries (Ray and Heyes, 2011). For example, if the instinct theory were correct—if imitation depended on a mechanism that computes topographic similarity—one would expect all perceptually opaque actions to be harder to imitate than perceptually transparent actions and, therefore, the imitation of opaque actions to emerge uniformly late in development. But this is not the pattern that has been found. Imitation of mouth opening and lip smacking begins to appear in Western infants at six to eight months of age (Kaye and Marcus, 1978; Piaget, 1962), whereas ear touching is not imitated until a year later (Uzgiris, 1972). This pattern suggests that, even among perceptually opaque actions, the age at which a particular action begins to be imitated depends on the extent to which infants have had the opportunity to learn a matching vertical association for that particular action. Matching vertical associations for mouth opening and lip smacking are established relatively early because, in contrast with ear touching, these actions are common targets for imitation of infants by adults (O'Toole and Dubin, 1968).

Homo imitans

Objection: Virtually all extant animals are capable of associative learning, but very few can imitate, and none are such skilled and prodigious imitators as humans. Therefore, imitation could not depend on associative learning.

The premise of this objection isn't quite right. After many years in which it was thought that very few nonhuman animals, perhaps only apes, were capable of imitation, it is now clear that a range of birds and primates can copy the topography of some body movements (Zentall, 2006). Indeed, there is some evidence that monkeys can copy precisely the topography of hand movement, that is, movements that allow matching vertical associations to be established by self-observation (Voelkl and Huber, 2007). However, the main thrust of the premise is sound. Humans really are prodigious imitators. We can imitate a much wider range of actions—especially novel, opaque actions—than any other animals. In this sense, we are "Homo imitans" (Meltzoff, 1988). The real problem with this objection is that there is a word missing: the taxonomic generality of associative learning, combined with the uniqueness of human imitative skill, indicates that imitation could not depend *exclusively* on associative learning. But when the missing word is inserted, the objection loses its force.

The cognitive gadget, ASL model of imitation in no way suggests that imitation depends exclusively on associative learning. Rather, it suggests that certain kinds of socio-cultural experience—for example, synchronous action, being imitated, and interacting with mirrors—are of overriding importance in the development of imitation. Only when this kind of experience is supplied by the developmental environment can associative learning build a substantial repertoire of matching vertical associations that solve the correspondence problem. And, although they have associative learning "on the inside," other animals lack these resources "on the outside." They do not have optical mirrors; action words to provide acquired equivalence experience; extended periods of development in which, at least in some cultures, adults regularly imitate infants; or the kinds of rituals, drills, and games—often involving music and dance—which provide humans with rich opportunities to see and do the same actions contingently (Heyes, 2016e).

Intentionality

Objection: Imitation is often done for a purpose. Sometimes people imitate because they want to fit in with a social group, curry favor, or acquire a skill. This would not be possible if imitation were based on learned associations because associations operate automatically, regardless of the person's intentions.

This challenge confuses learned associations with old-fashioned reflexes. The ASL model is entirely compatible with the duality of imitation, the fact that it is sometimes automatic and sometimes goal-directed, because it is built on contemporary associative learning theory, not on stimulus–response (S–R) behaviorism (Dickinson, 1980). S–R behaviorism suggested that all learning involves the formation of associations between stimuli and responses, such that, whenever the stimulus is encountered, the response is produced. In contrast, contemporary associative learning theory recognizes that S–R learning makes overt production of the response likely but not inevitable when the stimulus is encountered, and that associative learning can forge connections between representations of stimuli (S–S), and between representations of actions and their outcomes (R–S), as well as S–R links. Consequently, although the cognitive gadget view suggests that matching vertical associations are formed through associative learning, it does not imply that observation of an action for which the observer has a matching vertical association will always and everywhere ignite an overt imitative response. Rather, once a vertical association is in place, observation of an action will activate a corresponding motor representation, producing a conscious or unconscious "urge" to produce the action, which is detectable using neurophysiological measures of mirror activity. But this urge can be inhibited or facilitated according to what the observer knows about their situation and the likely consequences of performing the action (Heyes, 2011). So, when I see you wave, I have an impulse to wave back, but I can stop myself if I think you were waving at the person behind me, and I can amplify the impulse if I know you to be short-sighted.

Overimitation

Objection: Compared with chimpanzees, children "overimitate"—they copy silly, extraneous features of action in addition to the features that get results. This overimitation shows that humans have an inborn, species-specific proclivity to imitate which the cognitive gadget theory does not recognize.

When imitating retrieval of a toy from a box, three- and four-year-old children do not only release the latches and open the doors impeding their access to the prize. They also copy the model's extraneous actions—such as tapping the box with a wand—and show this "overimitation" even when they are able to tell the experimenter which actions are "silly" and which are necessary (Lyons, Young, and Keil, 2007). In this classic study of overimitation, and many others, it is not clear whether children are copying relations among body parts—what is called "imitation" in this book, and in cognitive science more generally—or whether they are reproducing relations among inanimate objects (for example, between the box and the wand), which is sometimes called "emulation." But this imitation-emulation distinction is not crucial because it is likely to be both. Children almost certainly imitate and emulate much more than chimpanzees, and much more than is necessary for them to fulfil their desires for inanimate things, such as toys, sweets, and stickers (Whiten, McGuigan, Marshall-Pescini, and Hopper, 2009).

The puzzle is why researchers have leapt to the conclusion that overimitation is due to a uniquely human cognitive instinct, whether it is called "social motivation," "shared intentionality" (Tomasello, 2014), or an "evolved heuristic" (Whiten et al., 2009). This is puzzling because there is plenty of evidence that: (1) adults reward children for imitating, with nods and smiles and by imitating them in turn (Grusec and Abramovitch, 1982); (2) rewarding children for imitating actions makes them more likely in the future not only to imitate those actions, but also to imitate other, physically similar actions (Baer and Sherman, 1964; Garcia, Baer, and Firestone, 1971; Young, Krantz, McClannahan, and Poulson, 1994); and (3) in the laboratory, rewarding chimpanzees for imitation has the same effects (Whiten and Custance, 1996). These findings make it plausible that children overimitate because, ever since they were born, they have been rewarded for imitation. Via plain old reinforcement learning, they have discovered that imitation tends to be followed by goodies, and it is the conscious or unconscious expectation of those goodies that makes them overimitate.

To be clear, this proposal relates not to ability, but to motivation. The cognitive gadget view of imitation suggests that both ability and motivation depend on learning, but on different kinds of learning. Matching vertical associations underpin the ability to imitate, and their development typically does not involve reward. I will acquire a matching vertical association for frowning if I receive correlated experience of seeing and doing frowning. It is not necessary for frowning or seeing frowning to be rewarding in itself, or for seeing and doing the action to be followed by reward. Once the ability to imitate is in place, we need to explain why it is used a lot or a little, in some circumstances and not in others; why the urge to imitate is sometimes inhibited and sometimes expressed. This, the motivation to imitate, and particularly to overimitate, depends on reward learning—on the discovery through experience that imitation of some actions, in some contexts, tends to have happy consequences, and the imitation of other actions, in other contexts, does not.

The distinction between ability and motivation, although important, is often elided in discussions of overimitation, and of "social motivation" and "shared intentionality" more generally. These concepts are candidates to explain why humans *want* to imitate, not how we are *able* to imitate. Unless we assume that wanting something is enough to make it happen, social motivation and shared intentionality do not address the correspondence problem.[3]

So, the occurrence of overimitation is not a problem for the cognitive gadget view of imitation for two reasons. First, overimitation raises questions about motivation—for example, the motivation to be accepted by one's desired in-group—rather than ability, and the ASL model is focused on the ability to imitate; it is focused on how we solve the correspondence problem. Second, unless or until studies show that overimitation is due to a genetically inherited motivation to imitate (or to share intentions, or to be like others), rather than to domain-general reward learning, overimitation is consistent with the cognitive gadgets theory more generally. It fits a picture in which the genetic starter kit for uniquely human mentality consists of Small Ordinary, rather than Big Special, components (Chapter 3).

What's the Use?

Objection: The cognitive gadget theory focuses on copying of configural body movements—how parts of the body move relative to one another, rather than in relation to objects. That kind of copying could not fulfil the primary function of imitation—high fidelity cultural inheritance of tool-use and other object-directed skills.

It is certainly true that the cognitive gadget theory focuses more explicitly on body movements than other theories that link imitation with cultural evolution. For example, Tomasello and his collaborators have characterized imitation as copying the "means" rather than the "end" of an action, and as "process copying" rather than "product copying" (Tennie, Call, and Tomasello, 2009). They have not said, in so many words, that to copy "means" or "process" necessarily involves copying body movements. But I don't think there is any fundamental disagreement here. The ASL model is rooted in sub-personal cognitive science and is therefore concerned with the correspondence problem. It spells out what other theories merely imply. Consequently, there is no reason to doubt that the mechanisms proposed by the ASL model—matching vertical associations gearing perceptual sequence learning to motor sequence learning—could support means or process copying, thereby contributing to high fidelity cultural inheritance of object-directed skills. For example, if the most efficient way for a novice to learn how to make a stone tool involves encoding and reproduction of the angles between the expert tool-maker's shoulder, elbow, and wrist joints as he strikes the core with the hammerstone, that encoding and reproduction could be done by ASL mechanisms.

However, because the ASL model is more explicit than any other about what is copied in the course of imitation—the topography of body movements—it raises the possibility that evolutionists have overlooked the most important function of imitation: high fidelity cultural inheritance not of object-directed actions, but of communicative and gestural skills (Heyes, 2013). These skills are rarely considered in discussions of human evolution, but their importance in defining groups and promoting cooperation is recognized in anthropology and the humanities (Corbeill, 2004). They include the sequences of body movements that enable group members to communicate without words and, thereby, to coordinate their activities when words are absent (for example, when the message is ineffable, and before language co-evolved), and when words are dangerous (for example, when a group is stalking prey). They also include the sequences of body movements, such as those involved in ritualistic dancing, that enable group members to bond—to achieve the states of trust and commitment required for cooperative action—through the expression of common religious beliefs, and the sharing of heightened states of arousal (Tarr et al., 2015; Whitehouse, 2004).

Thus, imitation enables the cultural inheritance of gestural skills that support cooperative action by communicating information, promoting trust and commitment, and indicating who is (and who is not) part of the cooperative group.

Some of the social practices that foster the development of human imitation may have been culturally selected for that "purpose." For example, groups with synchronous activities (rituals, drills, and games) that encouraged the development of a larger repertoire of matching vertical associations and, therefore, superior capacity to imitate, may have been more likely to prosper—to have more descendants—than groups with synchronous activities supporting the development of a smaller repertoire (see Chapter 9).

CONCLUSION

Imitation is the longest serving category of cultural learning. Scientists have been claiming for more than a century that imitation is a form of social learning specialized for cultural inheritance. In this chapter, I have embraced that view but challenged the assumption that imitation is a cognitive instinct. The ASL model, for which there is empirical support both from training studies and experiments showing that imitation has signature limits, not only suggests that imitation is a cognitive gadget, but provides a picture of how a new, specialized cognitive mechanism can be constructed by domain-general cognitive processes through social interaction. Specified types of sociocultural interaction (for example, synchronous action in dance, drills, and sports; being imitated by others; use of mirrors) yield correlated sensorimotor experience of seeing and doing the same action. This experience builds a repertoire of matching vertical associations. The matching vertical associations act like teeth on the wheel of perceptual sequence learning, gearing it to motor sequence learning. The resulting compound mechanism is a cognitive gadget capable of imitation learning.

The ASL model suggests that experience with optical mirrors can contribute to the development of the imitation mechanism. It does not say that mirror experience is necessary for the development of imitation. Nevertheless, some people still find the suggestion highly implausible, even shocking. Surely an artifact that emerged so recently in human history could not play a significant role in the development of a cognitive mechanism? I think this reaction is a measure of how accustomed we are to thinking about cognitive mechanisms as instincts, with deep roots in prehistory and the human genome. The next chapter continues to challenge this mental habit by suggesting that mindreading develops in much the same way as print reading, a cognitive capacity we know to be only five to six thousand years old.

7

MINDREADING

MINDREADING (ALSO KNOWN AS "MENTALIZING" and "theory of mind") is the process of ascribing mental states, thoughts and feelings, to oneself and others. Alongside imitation, mindreading is the psychological faculty most commonly put forward as a type of cultural learning. At first blush, this is a little surprising because, unlike imitation and other forms of social learning, mindreading is not in itself a process that enables social interaction to produce the kind of long-term changes in beliefs and behavior necessary for the cultural inheritance of knowledge and skills. In prototypical examples of mindreading, an agent works out what another agent is thinking or feeling *right now.* I would use mindreading to work out whether you think the ring tone we can both hear is coming from your bag or mine, and whether you are pleased or disappointed with the gift I have just given you. Answering these questions may well smooth the path of our interaction and enable me to predict your behavior in the short-term, but it does not amount to learning something of general and enduring significance about the world. Thus, mindreading is not, all by itself, a form of cultural inheritance. However, mindreading is important in relation to cultural evolution because it plays a crucial role in teaching.

Teaching is often defined in contrast with other kinds of social learning as a process in which one agent doesn't merely permit another to observe their behavior, but acts with the intention of producing an enduring change in the mental states—especially the knowledge states—of another agent (Byrne and Rapaport, 2011). On this view, the intention to change mental states, which presupposes mindreading, is what makes the first agent a "teacher" rather than just a "model," and the second a "pupil" rather than an "observer."

There is some debate about whether teaching should be defined such that mindreading is a necessary component (Caro and Hauser, 1992; Thornton and McAuliffe, 2012; Thornton and Raihani, 2008), but few would deny that mindreading is an important part of much human teaching. In principle, teachers could focus exclusively on what pupils are doing, not what they are thinking, and on how the pupil's behavior must change in order to meet a certain instrumental or social standard. However, assuming that behavioral competence (for example, skill in using tools, or navigating a particular set of social norms) really does depend on internal states, something like thoughts and feelings, teaching will be more effective if these internal states can be accurately represented by the teacher. Mindreading allows teachers to represent the extent and limits of a pupil's current knowledge and, thereby, to infer at each stage in the learning process what that particular pupil must be shown or told to overcome ignorance, correct false beliefs, and build his or her body of knowledge. And in a complementary way, mindreading by pupils enables them to isolate what it is that a teacher intends them to learn and, thereby, to focus their efforts on particular aspects of a to-be-learned skill.

Indeed, from a cognitive science perspective, mindreading is the best candidate for a "special ingredient" of teaching. Effective teaching certainly involves many other cognitive and motivational ingredients. For example, in addition to competence in the skill to be taught, good teachers are attentive to what their pupils are saying and doing, capable of distinguishing improvements in skill from backsliding, and socially tolerant enough to persist when progress is slow. But, compared with mindreading, none of these desiderata looks likely to depend on distinctively human cognitive processes that are specialized for social interaction. Social tolerance is a matter of temperament rather than cognition (see Chapter 3). Close attention to action and its effects, along with the capacity to distinguish improvement from backsliding, is present, to some degree, in a range of species and is important not only in teaching, or in social interaction more generally, but also when an agent learns or enacts a skill in social isolation. In contrast, mindreading appears to be distinctively human and to involve cognitive processes that are dedicated to processing input from other agents. Therefore, although it is unlikely that mindreading evolved (genetically and/or culturally) only for teaching, and likely that the upgrading of a range of other, domain- and taxon-general processes can contribute to making humans effective teachers, mindreading stands out as the most likely candidate for a human-specific cognitive adaptation for teaching.

So, where does mindreading come from? The dominant view suggests that mindreading is a cognitive instinct; that humans genetically inherit a specific predisposition to develop cognitive mechanisms representing mental states, and these mechanisms mature to much the same terminal state in a very broad range of developmental environments (Baillargeon, Scott, and He, 2010; Baron-Cohen, 1997; Leslie, 1987). A commonly held alternative view, "theory-theory," suggests that mindreading is learned via a quasi-scientific process

(Gopnik and Wellman, 2012). The developing child generates, through her own efforts, hypotheses about the mind and uses the behavior of others as a database for testing these hypotheses. To the extent that children end up with a species-typical or group-typical theory of mind, it is because, tested against the same database, their independently generated hypotheses converged on the same conclusions. In contrast with both the cognitive instinct and child scientist positions, I suggest that mindreading is a cognitive gadget. Children are taught about the mind by members of their social group, and the information that is culturally inherited in this way forms a conceptual structure enabling the ascription of mental states to the self and others. So, children are taught to read minds, and, once developed, mindreading contributes to teaching. Mindreaders can teach mindreading and a host of other culturally accumulated knowledge and skills.

The first section of this chapter outlines a number of similarities between mindreading and "print reading" or literacy. As discussed in Chapter 1, print reading is a proof of principle for cognitive gadgets. It is a distinctively human cognitive process that acquires its characteristic features through social interaction in the course of development. Literacy has not been around for long enough for the cognitive processes involved in print reading to have become genetically assimilated. Therefore, to the extent that mindreading is like print reading in the respects discussed in the first section—in ways that have made people think that mindreading is a cognitive instinct—we have at least cleared the ground for the possibility that mindreading is a cognitive gadget. The second section takes a more direct approach, reviewing evidence that the development of mindreading, like that of print reading, depends on teaching. The third deals with a potential problem for the view that mindreading is a cognitive gadget, discussing evidence that nonhuman animals, human infants, and adults-under-pressure are capable of "implicit" or "automatic" mindreading. The fourth elaborates and defends my "submentalizing" take on implicit mindreading.

MINDREADING AND PRINT READING

Like print reading, mindreading involves the derivation of meaning from signs. In print reading, the signs are usually marks on paper, and their meaning relates to objects and events in the world. In mindreading, the signs are facial expressions, body movements, and utterances—many of them conventional—and their meaning relates to the actor's mental states. But there are many deeper similarities between print reading and mindreading (Heyes and Frith, 2014). For example, both have regulative as well as interpretive aspects. As children are tutored in the writing conventions of their language, they learn not only to decode printed words—to relate them to spoken words and to meaning—but also to produce printed words that obey the conventions of their writing system so that they can be decoded by others. Research on mindreading has emphasized its interpretive role; the way that the attribution of mental states allows us to explain and predict behavior. However, "mental literacy" also has an important regulative role. Novice mindreaders learn not only that behavior *can* be, but that it *should* be, produced by rational interactions among beliefs and desires, and they are encouraged to make their own behavior obey these conventions. Adults ask children for reasons and mock (gently or otherwise) behavior that does not make belief-desire sense as "silly" or "crazy" (McGeer, 2007).

Mindreading is also like print reading in being slow to develop and cognitively demanding. The slow development of print reading is reflected in the fact that event-related potentials—scalp recorded electrical signals indicative of reading fluency—are not quite as fast even in sixteen-year-olds as in mature adults (Brem et al., 2006). Similarly, it is now known that, in Western cultures, the neural systems implicated in mindreading are among the last to reach maturity (Blakemore, 2008), and performance on tests of mindreading—including perspective-taking, emotion recognition, and detection of pretense and irony—continues to improve between adolescence and adulthood (Dumontheil, Apperly, and Blakemore, 2010; Vetter, Leipold, Kliegel, Phillips, and Altgassen, 2013).

Three other dimensions of comparison between mindreading and print reading are particularly interesting because they relate to features of mindreading that have been assumed to show genetic inheritance: neural specialization, developmental disorders, and cultural variation. Given that print reading is culturally, rather than genetically, inherited, if print reading is like mindreading with respect to these features, it suggests, at minimum, that the features do not provide evidence that mindreading is genetically inherited.

Neural specialization. Neuroimaging has shown that adults have cortical circuits specialized for mindreading. These circuits, which include the medial prefrontal cortex, temporo-parietal junction, and precuneus, are more active when people are thinking about mental states than when they are performing similar tasks that do not involve thinking about mental states (Van Overwalle, 2009). It is tempting to assume that this sort of cortical specialization is due to genetic evolution. However, similar specialization has been found in the case of print reading. For example, in literate adults, an area of the occipitotemporal cortex, known as the "visual word form area," is more active when people are viewing words than comparable non-word stimuli (Dehaene and Cohen, 2011).

Developmental disorders. Many people with autism spectrum conditions have a specific impairment in mindreading. Compared with IQ and language-matched controls, they have difficulty working out what others are thinking and feeling (Frith, 2001). Because there is a significant genetic contribution to the heritability of autism, this might suggest that neurocognitive mechanisms specialized for mindreading are usually genetically inherited, and that people with autism genetically inherit atypical versions of these mechanisms. But dyslexia reminds us that this is only one of many candidate explanations for the mindreading impairment associated with autism. Dyslexia is a genetically heritable developmental disorder that interferes with the acquisition of a culturally inherited skill, learning to read print (Paracchini, Scerri, and Monaco, 2007).

Cultural variation. At first glance, it seems that print reading is characterized by cross-cultural diversity and mindreading by cross-cultural commonality. In the print domain there is a rich diversity of scripts, and the size of the speech units that are mapped onto printed units varies from whole words in Kanji, to syllables in Japanese Kana, to phonemes in alphabetic writing systems (Dehaene, 2009). In contrast, in the mental domain, it is commonly assumed that the members of all cultures ascribe thoughts and feelings and understand these states to be related to behavior. However, recent research suggests that print reading is more, and mindreading is less, uniform across cultures than was previously supposed.

For example, featural analysis of 115 writing systems, contemporary and historical, has shown that most characters are formed by three strokes (Changizi, Zhang, Ye, and Shimojo, 2006). In a complementary way, psychological experiments are beginning to reveal significant cross-cultural variation in the development of mindreading. For example, children in Australia and the United States understand that different people can have different opinions ("diverse beliefs") before they understand that a person can be knowledgeable or ignorant of a particular fact ("knowledge access"). In contrast, children in China and Iran, where individualism and self-expression are less important, understand knowledge access before they understand diverse beliefs (Shahaeian, Peterson, Slaughter, and Wellman, 2011). Research of this kind supports earlier ethnographic data showing that cultures vary widely in the importance they assign to mental states, rather than social roles and situations, as causes of behavior, whether they take mental states to reside inside or outside the body, and in the extent to which they regard mental states as subject to natural rather than supernatural laws (Lillard, 1998). It also coheres with literary historical studies showing that, in Western cultures, ideas about the mind and mental disorders have changed radically since ancient times (Harris, 2013; Burrow, 1993).

These similarities between mindreading and print reading make clear that the main pillars supporting the cognitive instinct view of mindreading—evidence of neural specialization, of a specific impairment in autism, and of an invariant developmental sequence—are not structurally sound.

LEARNING TO READ MINDS

When more than a thousand pairs of five-year-old twins were given a comprehensive battery of mindreading tests, the correlation in performance within pairs was the same, 0.53, for non-identical twins with an average of 50 percent of their genes in common, and for identical twins who have all of their genes common. This indicates a "substantial shared environmental influence but negligible genetic influence on individual differences in theory of mind" (Hughes et al., 2005). In other words, it conflicts with the cognitive instinct view and suggests that learning plays a critical role in the development of mindreading.

However, by themselves, the twin data do not tell us what kind of learning is most important in the development of mindreading. The crucial learning could be of the type postulated by theory-theory, in which the child scientist tests hypotheses against behavioral data;

or of the kind envisaged by simulation theory (Goldman, 2006), in which the child works out what others are thinking and feeling through introspection, imagining herself in the same situation as another agent; or, as the cognitive gadget view suggests, the most important learning could be a consequence of teaching. Both theory-theory and simulation theory cast learning to read minds as an individualistic process. The target of learning, the thing-to-be-understood, is other people, but other people do not play an active role in advancing the learning process. They don't provide the hypotheses or perform the simulations. In contrast, the cognitive gadget view suggests that children are taught how to read minds in much the same way as they are taught how to read print: through scaffolding or "epistemic engineering" (Sterelny, 2012) and explicit instruction.

A "natural experiment" has provided compelling evidence that language-based teaching is crucial in the development of mindreading. The natural experiment was the emergence, in the 1970s, of a new sign language, NSL, among deaf people in Nicaragua. Pyers and Senghas (2009) compared false belief understanding in two adult cohorts of NSL users. They found that the first cohort, who learned NSL when it was still a rudimentary language, were less able to understand false beliefs than the second cohort, who learned NSL approximately ten years later, when it included many more signs for mental states. This suggests that, although the first cohort was ten years older—and therefore had much more time and opportunity to generate and test hypotheses (theory-theory), to reflect on their own mental states (simulation theory), and to observe the behavior of social partners—their understanding of false belief lagged behind that of second cohort. Lacking mental state vocabulary, they had not been able to receive instruction about the mind (Pyers and Senghas, 2009). Furthermore, the NSL findings are supported by a more recent study showing that children in Samoa, where it is considered improper to talk about mental states, develop an understanding of false belief at around eight years of age—four or five years later than children in Europe and North America (Mayer and Träuble, 2013).

In combination with other research involving typically and atypically developing children (Meristo, Hjelmquist, and Morgan, 2012; Slaughter and Peterson, 2012), the NSL and Samoan studies indicate that we learn about the mind through conversation about the mind. In principle, the appropriate conversational experience could come from "eavesdropping" or listening to what expert mindreaders say when they have no intention of teaching a novice. However, many studies suggest that experts—especially mothers—tailor their conversation so that it helps children to learn about the mind (Taumoepeau and Ruffman, 2006; 2008; Meins, 2012). For example, in conversation with their fifteen-month-olds, mothers make proportionally more references to desires and emotions than to thoughts and knowledge, and the frequency of desire / emotion talk selectively predicts the children's mindreading performance at twenty-four months. In contrast, at twenty-four months, mothers talk proportionally more about thoughts / knowledge than desires / emotions, and it is the frequency of references to thoughts / knowledge that predicts mindreading at thirty-three months (Taumoupeau and Ruffman, 2008). Desires / emotions are easier to understand than thoughts / knowledge because infants and children are constantly attempting to fulfil their desires, and emotions are often reflected in distinctive facial expressions. Therefore, these findings suggest that, just as novice print readers are introduced to "easy words" before "hard words," novice mindreaders are introduced to "easy states" before "hard states," and this epistemic engineering of the child's learning environment promotes the development of mindreading.

The development of mindreading is not only predicted by the frequency with which parents use mental state terms such as "think," "want," and "happy" in conversation with their children. It is also predicted by the frequency of parents' "causal-explanatory" statements about the mind, specifying relations between situations, mental states, and observable behavior. For example, "She is smiling because she is happy," or "She thinks the toy is in the green box because she didn't see it moved to the yellow box" (Slaughter and Peterson, 2012). In addition, Lohmann and Tomasello (2003) found that training sessions in which three- and four-year-olds conversed with an adult about deceptive objects (for example, a pen that looks like a flower) were especially effective in improving false belief understanding when the adult used sentential complement syntax—sentences such as "What do you think it is?" and "You say it is a flower," which take a full clause as their object complement. Research on causal-explanatory statements and sentential complements (de Villiers and de Villiers, 2012) suggests that conversation about the mind does not merely teach children labels for mental states. It also teaches them mental state concepts—what it is to "think" or to "feel" something, to be "happy" or "doubtful"—and gives them a format in which to represent these concepts. In other words, children culturally inherit from their parents and other mindreading experts (O'Brien, Slaughter, and Peterson, 2011) mechanisms that are specialized for the representation of mental states.

Thus, natural experiments, observational studies, and traditional experiments provide evidence that mindreading is taught in much the same way as print reading, through scaffolding and explicit instruction. Interestingly, very little of this empirical work was conducted in the United States, which may help to explain why it has received relatively little attention. Researchers in the land of the free, which dominates psychology as it does many other scientific disciplines, generally favor individualist conceptions of mindreading, in which the capacity to read minds is owned by virtue of genetic endowment and/or earned through the hard labor of hypothesis testing. In contrast, the cognitive gadget view of mindreading casts it as a collective achievement. Each culture has a theory of mind based on pooled experience, and each individual within the culture has a variant of the theory constructed in the course of development through active, cooperative interactions with other members of the group.[1]

IMPLICIT MINDREADING

Although there is a substantial body of evidence supporting it, the cognitive gadget view of mindreading also has a potential problem. Studies of infants now suggest that the development of mindreading is very different from that of print reading. Western children do not typically acquire the skill of print reading until they are five or six years old, but infants as young as seven months old seem to be capable of mindreading (Kovács, Téglás, and Endress, 2010). The most striking results suggest that infants ascribe false beliefs to agents; they understand that an agent can believe something that does not match the physical reality of the situation. For example, in a typical experiment on false belief ascription (Onishi and Baillargeon, 2005), infants first watch an adult placing a toy in one of two boxes, green rather than yellow, and subsequently reach repeatedly for that box. Then they see the toy moved from the green to the yellow box when the adult is present (true belief condition) or absent (false belief condition). In the final phase, infants in the true belief condition show more surprise, measured by looking time, when the adult reaches for the green box rather than the yellow box, while infants in the false belief condition are more surprised when she reaches for the yellow box. In both cases the infant "expects" a reach towards the box where the adult believes the toy to be located.

Research of this kind, which infers the occurrence of mindreading from nonverbal behavior such as looking time, is said to provide evidence of "implicit" mindreading. Implicit mindreading can be interpreted in three ways. The "continuity" interpretation suggests that it is controlled by the same, specialized cognitive mechanisms that mediate mindreading in adults (Baillargeon et al., 2010). The "two-systems" interpretation suggests that implicit mindreading, found using nonverbal measures, and old-style mindreading now dubbed "explicit mindreading" or "explicit mentalizing," arise from different cognitive mechanisms. Both systems are specialized for thinking about mental states, but the implicit system develops early and tracks mental states in a fast and efficient way, while the explicit system develops later, operates more slowly, and makes heavier demands on executive functions such as working memory and inhibitory control (Apperly, 2010; Perner, 2010; see Chapter 3 for a general introduction to "dual-" or "two-systems" models of cognition and discussion of executive function). In common with the two-systems model, the "submentalizing" interpretation assumes that explicit mindreading depends on specialized mechanisms that are late developing, slow, and effortful. However, the submentalizing view suggests that implicit mindreading depends on domain-general cognitive mechanisms, rather than mechanisms that are specialized for thinking about mental states (Heyes, 2014a; 2014b; 2015).

The case of autism suggests that implicit and explicit mindreading depend on different mechanisms since explicit mindreading can be achieved in spite of continuing problems with implicit mindreading (Senju, Southgate, White, and Frith, 2009). Further evidence for this dissociation is found in studies of neurotypical adults. In tasks where adults make verbal judgments about other people's thoughts and feelings (explicit mindreading), judgment accuracy is impaired by concurrent performance of an executive function task (Bull, Phillips, and Conway, 2008). In contrast, concurrent demands on executive function do not interfere with implicit mindreading (Qureshi, Apperly, and Samson, 2010). These results—indicating that explicit mindreading does, and implicit mindreading does not, depend on executive processes—are hard to reconcile with the continuity hypothesis but compatible with the two-systems and submentalizing interpretations of implicit mindreading.

The two-systems interpretation suggests that implicit mindreading depends on cognitive processes specialized for the representation of mental states. In contrast, the submentalizing interpretation suggests that implicit mindreading depends on domain-general mechanisms of attention, learning, and memory; processes that are genetically inherited but which did not evolve for, and are not dedicated to, mindreading or, more generally, the prediction of behavior (Chapter 3). I favor the submentalizing interpretation for two complementary reasons. First, we do not yet know what dedicated processes of implicit mindreading would look like. Butterfill and Apperly (2013) have made a bold and interesting attempt to specify how such processes work, but it is not yet clear whether what they describe as "minimal theory of mind" could produce the range of behavioral effects known as implicit mindreading *and* constitute a system that represents mental states, rather than purely observable, behavioral states (Butterfill, Apperly, Rakoczy, Spaulding, and Zawidzki, 2013). Second, having searched the literature on implicit mindreading—not only in infants (Heyes, 2014b), but in nonhuman animals (Heyes, 2015) and adults under time pressure (Heyes, 2014a)—I have been unable to find a case of implicit mindreading that is not explicable by domain-general processes well known to cognitive science.

For example, in a particularly strong study of implicit mindreading in infants (Southgate, Senju, and Csibra, 2007), measuring anticipatory eye movements, twenty-five-month-olds watched videos in which an adult's head was visible above two windows, and below each window was an opaque box (see Figure 7.1). At the beginning of each "familiarization trial," a puppet appeared at the bottom of the screen and placed a ball in one of the two boxes; in the left box in the first familiarization trial (Figure 7.1A), and in the right box in the second (Figure 7.1B). When the puppet disappeared, both windows were illuminated, a tone sounded, and, a second or two later, a hand came through the window above the box containing the ball and retrieved it from the box. Then the infant saw one of two "belief induction" sequences, False Belief 1 (Figure 7.1C) or False Belief 2 (Figure 7.1D), in which the puppet: (1) placed the ball in the left box, (2) moved the ball from the left to the right box, and (3) retrieved the ball from the right box and removed it from the scene. At a certain point in this sequence—after the puppet moved the ball from the left to the right box (2) in False Belief 1, and after the puppet deposited the ball in the left box (1) in False Belief 2—a bell began to ring, and the agent turned her head away. After (3), once the puppet had disappeared, the test trial started; the windows were illuminated, the tone sounded, and the researchers recorded the infant's eye movements. They found that the infants in the False Belief 1 condition made their first eye movement to the right window, whereas infants in the False Belief 2 condition made their first eye movement to the left window.

As adults watching these videos, or reading a description of the experiment, it is natural for us to do plenty of mindreading. We interpret the familiarization trials as indicating that the agent will reach for the box in which she *believes* the ball to be hidden; that the head turning is a sign that the agent did not *see,* and therefore did not *know* about, the movements of the ball; and that the infants' eye movements are evidence that they expected the agent to reach for the box in which the agent (falsely) *believed* the ball to be hidden. But there is another likely explanation for the infants' looking behavior: distraction or,

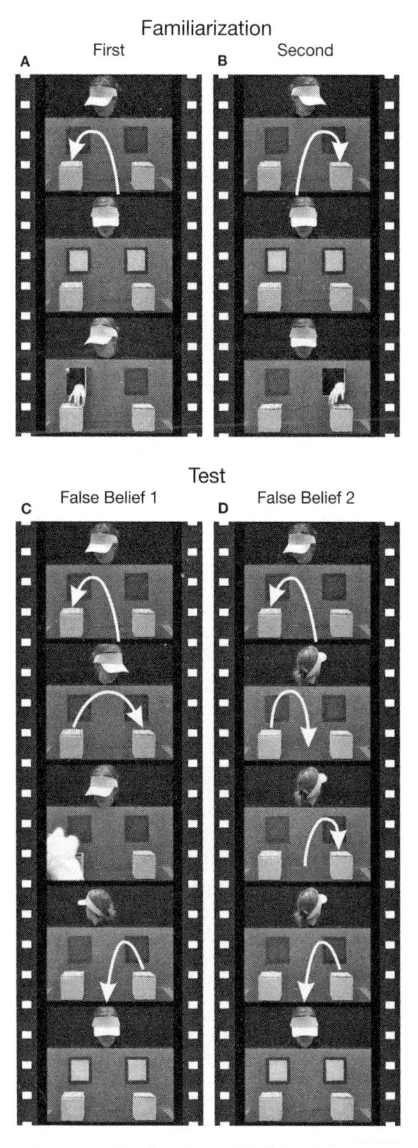

7.1 Stills from the videos used by Southgate and colleagues to test for the ascription of false beliefs by twenty-five-month-old infants. (Reprinted with permission from Southgate, Senju, and Csibra, 2007.)

more specifically, "inattentional blindness" (Mack and Rock, 1998). Like adults who fail to notice a gorilla in a group of humans because they are focused on a counting task (Simons and Chabris, 1999), the infants may have failed to notice the ball movements while the bell was ringing because their attention was focused on the agent's turned head and the area to which she was looking. Thus, the infants made their first eye movements to the last location at which they, the infants, had seen the ball: to the right box in False Belief 1, the condition in which the agent turned after the ball was transferred from left to right, and the left box in False Belief 2, the condition in which the agent turned after the ball was initially placed in that left box.[2]

The evidence of implicit mindreading in nonhuman apes could also be due to domain-general processing (Heyes, 2015). For example, Krupenye and colleagues used a procedure modeled on the study by Southgate and colleagues (2007), described above, to test for false belief ascription in chimpanzees, bonobos, and orangutans (Krupenye, Kano, Hirata, Call, and Tomasello, 2016). In their experiment, the agent, rather than turn away, left the scene during critical movements of the target object. This was an advantage in that it allowed the apes, unlike the infants, to view the object movements without distraction. However, it had the unfortunate side effect of making the agent, who wore a bright green shirt, into an ideal retrieval cue, that is, a stimulus that triggers recall of an event because it was prominent when the event was encoded. The apes may have *seen* the target object at each location but, guided in the test by the pattern established during familiarization trials, retrieved from memory the location at which it last appeared with the bright green stimulus (Heyes, 2017b).

Mindreading can be distinguished from domain-general processing using inanimate control procedures, in which social stimuli are replaced by non-social stimuli with similar low-level perceptual features, for example, colors, shapes, and dimensions. This technique has not been used with infants or apes, but, applied to adult humans, it has provided evidence that implicit mindreading is due to domain-general processing (Catmur, Santiesteban, Conway, Heyes, and Bird, 2016; Santiesteban, Catmur, Hopkins, Bird, and Heyes, 2014). This research was based on the "dot perspective task" in which people are presented with images like those on the left of Figure 7.2 and asked to make quick judgments about the number of dots that they, the experimental participant, can see. Providing evidence of implicit mindreading, participants typically make faster judgments when all dots are in front of the central figure, the "avatar" (Figure 7.2A), than when some dots are behind the avatar (Figure 7.2B; see Samson, Apperly, Braithwaite, Andrews, and Bodley Scott, 2010). Using an inanimate control procedure, Santiesteban and

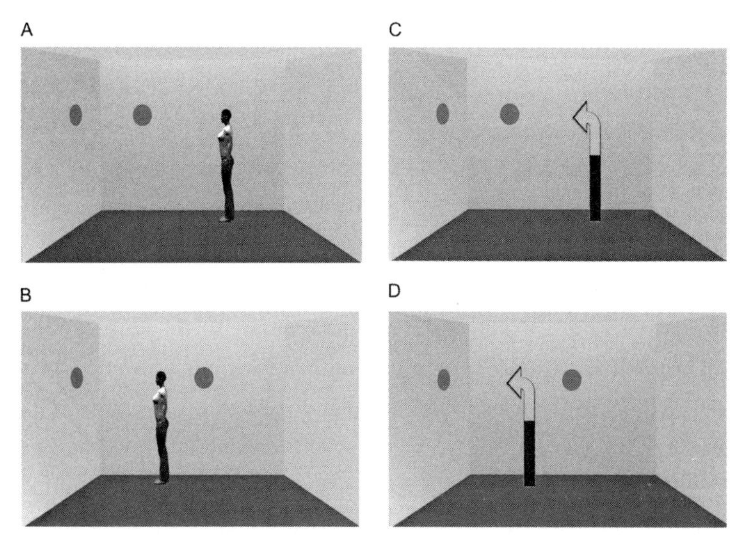

7.2 Examples of the stimuli used by Santiesteban and colleagues to test implicit mindreading in neurotypical adults. (Reprinted with the permission of the American Psychological Association from Santiesteban, Catmur, Hopkins, Bird, and Heyes, 2014.)

colleagues found that this effect also occurs when the central object is an arrow—an inanimate object rather than a human figure (Figures 7.2C and 7.2D)—suggesting that the effect is due to the domain-general mechanisms that mediate automatic attentional orienting (Tipples, 2008).[3]

These three examples—involving infants, chimpanzees, and human adults—support the submentalizing interpretation of implicit mindreading; the view that it is due not to fast-and-efficient processes that represent mental states, but to the operation of low-level, domain-general cognitive mechanisms (for reviews, see Heyes, 2014a; 2014b; 2015).

SUBMENTALIZING

There is a range of domain-general perceptual, attentional, learning, and memory processes known to cognitive science through research in which humans and other animals respond to inanimate stimuli (Eysenck and Keane, 2015). These processes differ from mindreading in that they parse the world into relatively low-level features, such as colors, shapes, and movements, rather than constructing a narrative in which agents act on objects for reasons. The submentalizing interpretation of implicit mindreading suggests that when subjects—infants, nonhuman animals, and adults under time pressure—are presented with animate stimuli, these domain-general mechanisms can simulate the effects of mindreading. For example, they can make accurate predictions about observed behavior. When they perform a job traditionally associated with mindreading, the domain-general mechanisms can be described as "submentalizing," but they were not made by genetic or cultural evolution for the purpose of tracking others' behavior.

The submentalizing interpretation of implicit mindreading is subtly but significantly different from the two-systems view (Butterfill and Apperly, 2013). They are both dual-process theories of the kind discussed in Chapter 3 (James, 1890). Both assume that human cognition involves two kinds of mechanisms: System 1 processes that are fast, automatic, parallel, and based on information derived from genetic inheritance and from domain-general processes of learning (for example, motor skills); and System 2 processes that are slow, effortful, serial, and based on information both from System 1 and generated by its own activity. Furthermore, they both assume that implicit mindreading is mediated by System 1, whereas explicit mindreading depends on System 2, and that the System 1 processes mediating implicit mindreading are genetically inherited. However, the two-systems view assumes that implicit mindreading depends on a System 1 process *dedicated* to mindreading, whereas the submentalizing account suggests that implicit mindreading is due to a wide range of System 1 processes, none of which represents mental states or is specialized for social interaction. There is experimental evidence that the range includes processes responsible not only for associative learning, but involuntary attentional orienting, spatial coding of responses locations, object-centered coding of stimulus locations, retroactive interference, cued retrieval, and distraction (Heyes, 2014a). But in principle, any domain-general System 1 process could contribute to tracking the behavior of other agents.

I gave "submentalizing" its name, rather than calling it "quasi-mentalizing" or "fake mentalizing," because I believe the System 1 processes responsible for implicit mentalizing do a lot more than mislead us into thinking that babies, chimps, and harried adults are representing mental states. These System 1 processes also provide substitutes and substrates for genuine, explicit mindreading (Heyes, 2014a).

As substitutes, submentalizing processes allow effective cooperation and competition without the labor of representing mental states. For example, studies of implicit mentalizing in adults suggest that we habitually code the location of our own actions relative not only to objects and our own bodies, but to the locations of other agents' actions (Dolk et al., 2011). The System 1 process responsible for this kind of spatial coding of response locations is likely to play a crucial role in allowing smooth coordination of action. If you and I are moving furniture or dancing together, our performance will be more efficient and graceful if we individually code our movements, and the locations of key objects and events, relative to one another's body as well as to our own. Indeed, if we are both doing this kind of spatial coding, via domain-general processes, it is not clear what we would gain—how the efficiency and grace of our movements would be enhanced—if we were also representing what the other sees, believes, and intends. Similarly, domain-general attentional processes, of the kind producing automatic attentional orienting (see infant and adult examples above), can bring the mental states of two agents into alignment without either agent representing that alignment. Regardless of whether I am representing your mental states, if I attend to the same location as you, we are more likely to have the same beliefs at the same time—to be thinking simultaneously that "the cliff edge is nearby," "there's a puddle on the floor," or "someone has just entered the room." Like spatial coding of actions, this kind of attentional process can provide a substitute for mindreading in many behavioral coordination tasks, giving partners' similar targets and priorities for action.

Submentalizing provides a substrate for genuine mindreading primarily by building an "action vocabulary." Here, again, the analogy with print reading may be helpful. Through associative learning, an infant listening to speech discovers statistical regularities that define components such as words and syllables (Jusczyk, 1999). This knowledge of components is a vital precursor for learning how to map sounds onto strings of letters. Likewise, there are detectable statistical regularities intrinsic to body movements, due partly to physical constraints created by gravity and the construction of mammalian bodies (Thurman and Lu, 2013). Infants start out parsing behavior only in terms of its low-level features, such as color, shape, and movement. But, through the detection of statistical regularities intrinsic to body movements, associative learning soon begins to build higher-level categories and to detect sequential dependences; to predict where a movement will end, and which types of movement tend to succeed one another (Falke-Yitter, Gredebäck, and von Hofsten, 2006; Frank, Vul, and Johnson, 2009). For example, it becomes possible for the infant to encode an observed body movement not merely as something we might render in words as "rounded-pinkish towards angular-green," but as "hand towards green box." The hand movement no longer comes as a surprise because associative learning has detected that hand movements tend to be preceded by head movements, or—an even more reliable relationship—by eye movements, in the same direction. The infant starts to have an expectation about what will happen when the hand reaches the box: it will disappear and then reappear with another object. The resulting action vocabulary provides an initial set of referents for mental state terms and for instructions about how mental states relate to behavior (Skerry, Carey, and Spelke, 2013). Through conversation about the mind (see previous section, "Learning to Read Minds"), children do not just learn labels for the high-level features of behavior they have discovered by associative learning. In addition, through the operation of System 2 processes, rather than associative learning, they discover yet more abstract features of behavior and develop a whole, new way of thinking about action—as the product of internal states that represent the world and each other, that is, explicit mindreading. But the substrate, or scaffold, for this linguistically mediated learning is the action vocabulary established by submentalizing.

The submentalizing hypothesis, and the cognitive gadget theory of mindreading more generally, assumes that "language comes first." Through language, children learn to read minds. Consequently, the cognitive gadget theory is deeply implausible for those who believe that "mindreading comes first," according to which the ascription of mental states is a precondition for linguistic communication (Bloom, 2001; Sperber and Wilson, 1995; Tomasello, 2003). The idea that mindreading comes first is inspired by the Gricean account of language (Grice, 1957), in which a communicative act (verbal or non-verbal) has linguistic meaning only if the agent performing the act, A, *intends* the receiver of the communication, B, to *recognize* that the act is performed by A with the *intention* of making B respond in a particular way. Thus, A's action must be rooted in a mental state which is about B's mental state, and the mental state ascribed to B by A must be about A's mental state. On the Gricean view, an intention to communicate is an intention to change others' mental states (Sperber and Wilson, 1995).

I am not qualified to take a view on the merits of the Gricean account of language, but I doubt for two reasons that it conflicts with the cognitive gadget, language-first view of mindreading. First, even enthusiastic Griceans do not see Grice as having offered a psychologically realistic account of what is happening whenever people talk to one another; he provided a "rational reconstruction" of language rather than a scientific theory (Sperber, 2000b). Second, Moore (2016; in press) has argued persuasively that Gricean communication can get off the ground, in evolutionary and developmental time, with minimal mindreading. All that is needed is "a basic understanding of others' purposive activities and desires, operating in conjunction with some tracking what others had or had not seen" (p. 19). Adhering to Moore's analysis, I think these requirements can be characterized in a yet leaner way, making clear that they can be met by submentalizing. To get started with language, a child must know something about action-outcome relationships (rather than "purposive activities and desires") and be able to keep track of what others have viewed (rather than "seen").

Supporters of the mindreading-first position suggest that, whether or not complex mindreading is required for Gricean communication, the evidence of implicit mentalizing in infants shows that humans are able to read minds long before they can have a conversation. In contrast, I have argued here and in the previous section that implicit mentalizing in infants, chimpanzees, and adults under time pressure is due to domain-general cognitive processes—to submentalizing, not mindreading.

I have argued for the submentalizing interpretation of implicit mindreading because I think it provides the best fit with currently available evidence. However, it should be noted that the cognitive gadget view of mindreading does not depend on the validity of the submentalizing interpretation. Both the two-systems and submentalizing hypotheses suggest that mindreading or "explicit mentalizing"—the kind that allows us to deliberate about mental states and to express our thoughts about mental states in words—is a cognitive gadget.

CONCLUSION

In summary: mindreading is like print reading in having regulative as well as interpretive aspects; in being cognitively demanding and slow to develop; and in being characterized by neural specialization, developmental disorders, and cultural variation. Furthermore, evidence from natural experiments, observational studies, and traditional experiments indicates that, like print reading, mindreading is learned through scaffolding and explicit instruction. It is culturally inherited; a cognitive gadget. Expert mindreaders communicate mental state concepts, and ways of representing those concepts, to novices. As the novices become expert, they pass on the knowledge and skill of mindreading to the next cultural generation.

The analogy between print reading and mindreading is not perfect. "Sound symbolism" shows that the relations between inscriptions and their corresponding speech sounds and referents often depend on features of the nervous system (Ozturk, Krehm, and Vouloumanos, 2013), but it is likely that these relations are more arbitrary than the relations between observable behavior and mental states. In this respect, numeracy may be a better analogue than literacy. However, mindreading is comparable to print reading not only in terms of its weak dependence on genetically inherited mechanisms and strong dependence on teaching, but also in the shape and size of the cultural legacy. Like print reading, mindreading mechanisms represent representational relations—between mental states, behavior, and events in the world—and allow the mindreader to regulate and interpret a virtually limitless range of mental contents. Consequently, along with print reading, mindreading is a special cognitive ingredient of teaching—a practice with enormous potential to enhance the fidelity of cultural inheritance. The next chapter focuses on the other powerful ingredient of much human teaching: language.

8

LANGUAGE

LANGUAGE IS OFTEN DESCRIBED AS A RUBICON, A shining threshold in the evolution of human cognition. Ancient legend in many cultures suggests that, once the language boundary was crossed, the minds and lives of our ancestors were forever transformed. Now capable of abstract thought and subtle communication, we became radically different from all other animals—more like gods than beasts.

In this chapter, I do not wish to challenge the importance of language. Abstract thought may be less decisive in human affairs than intellectuals would like to believe, and linguistic communication is far from the only significant channel of cultural inheritance, but there can be no doubt that language is a remarkable skill, one of the foremost faculties that make humans unique. What I want to explore is how language evolved, genetically or culturally, and whether its reputation as a god-like faculty has influenced scientific attempts to answer this question.

The previous three chapters also asked whether a particular cognitive faculty—selective social learning, imitation, mindreading—is a product of genetic or cultural evolution. In this chapter, I report for outsiders *as* an outsider. As a cognitive scientist with a special interest in social cognition, I have been involved in research on social learning, imitation, and mindreading for decades, but, before I began to write this book, I kept well clear of language. Starting as a psychology student around 1980, whenever I peeped over the fence at what was happening in the language sciences it struck me as alien. And it wasn't just the jargon; it was the odd character of the ground rules. It seemed that very little weight was given to the information one would expect to be crucial in explaining language. There were people doing fine research on animal communication, how children acquire language, and how adults are able to generate and understand speech at such incredible speed, but the fruits of their labors did not seem to be having much impact on how language was explained. The Big Questions were about "Universal Grammar," and they were being addressed not by studying minds, but by studying sentences. More puzzling still, although the hunt was for universals, most of the sentences were in English. So, each time after peeping over the fence, I would scuttle away, hoping that, next time I looked, it would all make more sense. Now, thirty-five years after I first started peeping, and after a lot of reading, it does make more sense. Consequently, this chapter is intended, in part, as a guide for the perplexed; for people outside the language sciences who, like me, have been wondering what has been going on over there.

The first section contrasts the genetic evolutionary account of the origins of language, rooted in the work of Chomsky, with a cultural evolutionary account based on the "constructivist" approach to language. The second discusses two phenomena—linguistic universals and critical periods for language development—which, although they are often assumed to be decisive, are neither here nor there; they do not favor the genetic over the cultural hypothesis, or vice versa. The third part focuses on recent research indicating that language processing is widely distributed across the brain; that speech production and comprehension depend on domain-general processes of sequence learning; and that children are shaped by other agents to use the language typical of their social group. This research does not show that the cultural evolutionary hypothesis is right, and the genetic evolutionary hypothesis is wrong. Indeed, as I argue in the fourth section, the character and history of the genetic account give reasons to doubt that it could be discredited by any information about how the mind actually works. However, as discussed in the final section, recent discoveries about neural localization, sequence learning, and social shaping make the cultural account of the origins of language fully plausible. For the perplexed, the nativist story is not nearly as compelling as it looked from over the fence, and the less alien alternative has come of age.

TWO GAMES IN TOWN

Genetic Evolution of Language

Technically, there is a family of genetic accounts of the evolution of language, also known as Universal Grammar or "nativist" theories. The members of this family are united in two respects. First, they all descend from the work of Chomsky in the late 1950s and early 1960s. Second, they all claim that language learning is made possible by genetically inherited information *about language*. No one doubts that a wide range of genetically inherited resources are needed for language learning, including memory capacity, domain-general perceptual and motor skills, and dedicated vocal apparatus. What is distinctive about genetic accounts is their commitment to the idea that the genetic resources supporting language acquisition include information about the abstract structure of language; about features of grammar that, they claim, all languages have in common.

Views about the content of this language-specific, genetically inherited information have changed radically over time, and there continues to be a wide range of opinion. For example, Chomsky (1957) originally suggested that humans genetically inherit knowledge of "transformations," rules specifying how new sentences (such as the passive "Bessie was attacked by Rudolph") can be generated from old ones (the active "Rudolph attacked Bessie"), and that the child, like a little scientist, uses these transformations to find out about grammar via domain-general inferential processes (Cowie, 2016). In contrast, Chomsky's later "principles and parameters" approach (Chomsky, 1981; 1988) cast language acquisition as a passive process of maturation or unfolding, like a chicken growing a wing, and suggested that our inborn linguistic knowledge is much more extensive. Rather than transformation rules, it consists of full-blown, crisply specified grammatical principles, and switches that are flicked by the linguistic environment—the sentences children hear as they grow up—to settings appropriate for a particular natural language (Chomsky, 1988: 61). Finally, in more recent work, Chomsky and his collaborators have cut Universal Grammar back to the bone (Berwick and Chomsky, 2015; Chomsky, 1995; Hauser, Chomsky, and Fitch, 2002). Their "minimal program" suggests that the only genetically inherited, language-specific information is that which enables an operation called "merge." This operation combines two syntactic objects (word-like mental elements, or compounds of such elements previously produced by merge) to make a new syntactic object. For example, the word-like elements "read" and "books" might be merged to generate the verb phrase "read books," and this new object could then be merged, in a hierarchical way, with other word-like elements and compounds to produce larger, more complex syntactic objects that can be used repeatedly and self-referentially within a sentence. In contemporary language science, some theorists support the minimal program (Boeckx, 2006), others have persisted with something very close to the principles and parameters approach (Crain, Goro, and Thornton, 2006), and still others, promoting "simpler syntax," have streamlined that approach, arguing that the inborn principles of grammar provide a guide for learning rather than a blueprint for maturation (Culicover and Jackendoff, 2005; Pinker and Jackendoff, 2005).

As part of their legacy from Chomsky's (1965) early work, genetic accounts of the evolution of language distinguish sharply between "competence" and "performance." Linguistic competence is defined by whatever a given theory takes to be the content of Universal Grammar. Competence is genetically inherited, the same in everyone, and embraces an unbounded number of grammatically correct sentences. The study of competence is the preserve of linguistics, which attempts to formulate maximally elegant and economical rules that apply to all natural languages, testing these rules—candidate components of Universal Grammar—against native speaker intuitions about what is and is not a grammatically correct utterance in their language. In contrast, the term "performance" was used by Chomsky to refer to everything relating to language that is not part of Universal Grammar (Lyons, 1977). This category includes all of the psychological processes involved in the ontogenetic development and online control of language; the processes through which utterances are produced and understood. It is assumed that, due to speed and capacity limitations, these processes result in speakers using a tiny fraction of the sentences licensed by their inborn knowledge of grammar. Compared with competence, performance is feeble thing, and it is the province of psychology, sociology, and anthropology,

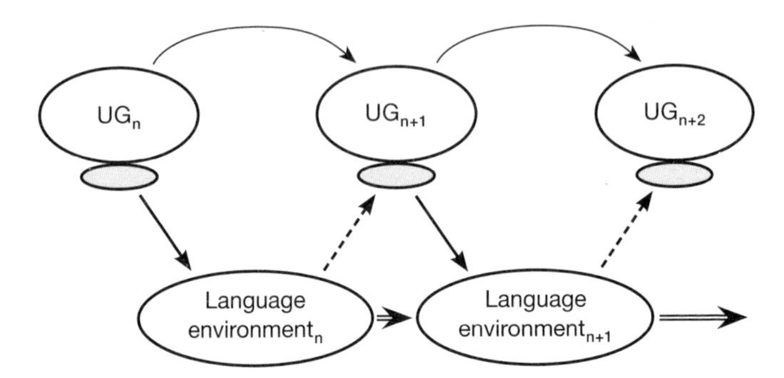

8.1 A schematic representation of gradualist genetic theories of language evolution. (Adapted with permission of The MIT Press from Christiansen and Chater, 2016.)

not of linguistics. Chomsky's competence / performance distinction entrenched a division of labor in which linguists typically study abstract, idealized properties of language using native speaker intuitions, while (other) language scientists draw on field observations and laboratory experiments to find out how, in the real world, language develops and is used for communication.

Figure 8.1, taken from Christiansen and Chater (2016: 9), represents the key features of contemporary "gradualist" genetic theories of the evolution of language (Culicover and Jackendoff, 2005; Pinker, 1994; Pinker and Bloom, 1990). These theories are developments of Chomsky's work on Universal Grammar. However, unlike Chomsky's own view, which treats the language faculty as having emerged recently and suddenly through chance mutation (Berwick and Chomsky, 2015; Chomsky, 1972; Hauser et al., 2002), gradualist theories are comparable with cultural accounts of the evolution of language in suggesting that language has been shaped by selection processes. The three ovals at the top of Figure 8.1, connected by curved arrows, represent successive forms of Universal Grammar (UG_n, UG_{n+1}, UG_{n+2}); that is, forms of the genetically inherited information about language that constitutes linguistic competence. The shaded ovals represent performance processes. These are responsible for the production and comprehension of sentences (oblique arrows) and are, therefore, driven and constrained by Universal Grammar; they generate the language environment (solid oblique arrows). This environment—the utterances that are actually spoken and heard—changes over time (double arrows) in a broadly cultural way. Within the constraints imposed by Universal Grammar, new ways of speaking emerge and spread through social learning, while others decline. Over long stretches of time, these changes in the language environment exert new selection pressures on the performance processes (dashed oblique arrows) and, through them, on the genes that encode Universal Grammar.

Cultural Evolution of Language

While the genetic theory of the evolution of language is rooted in the Universal Grammar or "structuralist" approach to linguistics, the alternative cultural theory is rooted in what is variously known as "constructivist," "cognitive," or "functional" linguistics (Bates and MacWhinney, 1982; 1989; Lakoff, 1990). In an excellent, short summary of the differences between the structuralist and functionalist approaches, Tomasello (1995) highlighted four contrasts. First, desiderata: structuralism seeks mathematical elegance, while functionalism values psychological plausibility. Second, fundamental distinctions: structuralism contrasts meaning with syntax, understood as the order of components in a sentence, while functionalism contrasts meaning with all properties of "linguistic symbols," including (when the symbols are sentences) the order of components as well as the particular words used and their inflections. Third, universals: structuralism attributes commonalities across languages to inborn language-specific information, whereas functionalism assumes that, insofar as linguistic universals exist, they arise from the fact that all humans have the same domain-general cognitive resources, communicative purposes, and channels of communication. Fourth, development: structuralism assumes that the ontogeny of language is the unfolding of a genetic blueprint, or that it is guided by psychological processes dedicated to language learning—a "language acquisition device." In contrast, functionalism assumes that language acquisition depends on domain-general cognitive processes guided by linguistic and nonlinguistic input from other agents.

At least two cultural evolutionary accounts of the origins of language have been developed from functional linguistics, one by Tomasello (2003) and the other by Christiansen and Chater (2016). I will focus on the latter because, building on Tomasello's earlier work, Christiansen and Chater's theory is more explicit and makes more extensive use of contemporary research in cognitive science. Figure 8.2 shows Christiansen and Chater's schematization of their theory. The three large ovals at the top, connected by curved arrows, represent successive states of the system that is evolving. This system comprises not a Universal Grammar (Figure 8.1), but a set of innumerable and interrelated "constructions" or "processing events" (P_h–P_k)

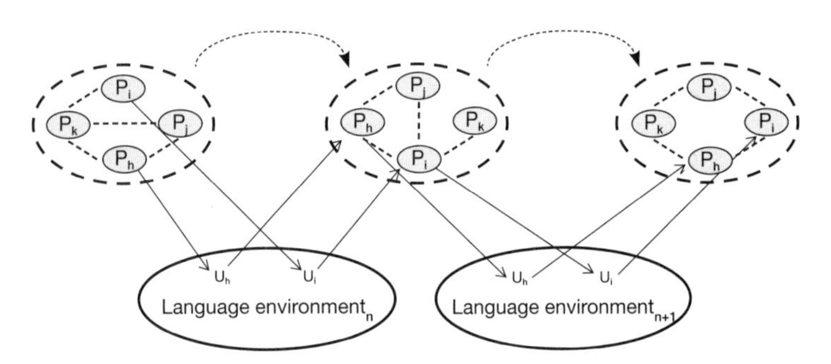

8.2 A schematic representation of Christiansen and Chater's cultural theory of language evolution. (Adapted with permission of The MIT Press from Christiansen and Chater, 2016.)

occurring within a speech community (dashed line defining each oval). Processing events can be attempts to understand utterances in the language environment (upward oblique arrows) or to produce utterances that contribute to the language environment (downward oblique arrows). Knowledge of language is acquired, over many years, through innumerable processing events—interactions with the language environment—and via generalizations across processing events (dashed lines between Ps). This knowledge is encoded by domain-general memory and sequence learning processes, which predate the emergence of language; the cognitive processes of language acquisition evolved genetically to fulfil nonlinguistic functions. Indeed, in Christiansen and Chater's account, the constraints imposed by domain-general mechanisms are the selective environment for the cultural evolution of language. In other words, language is "shaped by the brain" (Christiansen and Chater, 2008). Constructions / processing events that are easier to handle by domain-general mechanisms are more likely to yield comprehensible utterances and, therefore, to proliferate in the speech community. Over time, this selection process changes the population of processing events—that is, the language of a speech community—making it better adapted to domain-general cognitive processes and, therefore, more expressive.[1]

Reflection on Chomsky's famous poverty of the stimulus argument makes it easier to see the essential differences between the genetic and cultural accounts of the evolution of language. Chomsky rarely presented his poverty argument in a general form but, simplifying Cowie (2016), it can be summarized:

Premises

1. Mastery of a language requires knowledge of Universal Grammar.

2. In order to learn Universal Grammar, children would need access to certain kinds of data.
3. These data are not available in children's linguistic environments.

Conclusions

1. Universal Grammar could not be learned.
2. Universal Grammar is genetically inherited and / or innately known.

Building on functionalist or constructivist linguistics, the cultural account of the evolution of language challenges all three premises of the poverty argument. It claims that: (1) mastery of language requires knowledge of constructions, not of Universal Grammar; (2) knowledge of constructions is acquired by domain-general processes of sequence learning; and (3) the data required for this learning *are* available in children's linguistic environments. The third section of this chapter surveys some recent research supporting these three claims, but first let us take a quick look at two long-standing fields of research on language—universals and critical periods—that can easily be mistaken for having yielded decisive evidence in favor of the cultural account.

LONG STORIES

Universals

It has been estimated that there have been 500,000 human languages (Pagel, 2000). Five to eight thousand languages are spoken in the world today. Of these extant languages, only about 10 percent have been documented in detail. Nonetheless, research in comparative linguistics has revealed breathtaking diversity in phonetics (including sign language), morphology, semantics, and syntax (Evans and Levinson, 2009). For example, there are languages without adverbs, languages without adjectives, and, some evidence suggests, languages without a basic distinction between nouns and verbs (Jelinek, 1995). Furthermore, some languages have major word classes in addition to the Big Four—nouns, verbs, adjectives, and adverbs—that feature in Indo-European languages. The members of one of these additional classes, ideophones, typically encode in an integrative way the cross-modal perceptual properties of an event, such as the Mundari words "ribuy-tibuy," referring to the "sound, sight, or motion of a fat person's buttocks rubbing together as they walk," and "rawa-dawa," indicating "the sensation of suddenly realizing you can do something reprehensible, and no-one is there to witness it" (Osada, 1992). More generally, Evans and Levinson (2009) have identified evidence from comparative linguistics challenging each of the leading claims about what all languages have in common. In addition to the proposal that all languages have the Big Four word forms, these include claims that every language has: (1) major phrasal categories, (2) phrase structure rules, (3) rules of linear order, (4) verb affixes, (5) auxiliaries, (6) anaphoric elements, (7) numerals, and (8) "wh-movement," a property allowing interrogative words, such as *who, what,* and *which,* to appear at the beginning of sentences.

Thus, the picture that emerges from comparative linguistics is one of rich diversity among languages. A wealth of evidence challenges claims of the form, "all languages have X." This picture dispels the impression that there is solid evidence of what most people would understand by "linguistic universals": features that all languages have in common, in addition to those defining what it is to be a language (for example, discreteness, arbitrariness, and productivity).

Surprisingly, however, given that the genetic account of the evolution of language is founded on "Universal Grammar," the lack of features common to all languages does not count in favor of the cultural over the genetic account. There are two reasons for this. First, the cultural evolutionary account does not predict diversity on all dimensions, or, to put it another way, it does not forbid universals. It could turn out that all languages share certain features because they all descend culturally from the same Ur language (Cavalli-Sforza, 1997), or because the cultural evolution of every human language has been constrained by the same domain-general cognitive resources, communicative purposes, and channels of communication (Tomasello, 1995). Thus, on a straightforward understanding of "linguistic universals," they could be due to culturally inherited information and/or to domain-general genetically inherited information. They need not be due to inborn knowledge about language. Second, supporters of the genetic account and, more generally, of Universal Grammar, do not use the term "linguistic universals" in a straightforward way. For them, the linguistic universals that matter are *by definition* features of Universal Grammar; they are made possible by genetically inherited information about language. Or, to put it another way, they are features of a genetically inherited language of thought, not of natural languages (Berwick and Chomsky, 2015). A "universal" in this sense need not be present in all or even most natural languages, and a feature that was found to be present in all languages would not necessarily be a universal (Boeckx, 2006; Chomsky, 1965; Pinker and Jackendoff, 2005).

This view of what constitutes a linguistic universal makes it impossible to test for linguistic universals in any way that a cognitive scientist would recognize. The most vivid example of immunity to empirical testing relates to syntactic recursion, the one feature that Chomsky has consistently held to be the core component of Universal Grammar. Data from comparative linguistics have shown that recursion is extremely limited in several languages, and, controversially, some data suggest it is absent in at least one: the Amazonian language Pirahã (Everett, 2005). Nevertheless, supporters of the genetic account of the evolution of language are undisturbed by this (Fitch, Hauser, and Chomsky, 2005). They do not have a list of features that they agree are linguistic universals, or a set of agreed criteria for universality, but supporters of the genetic account are united in denying that the universality of a feature depends on its distribution across natural languages.

Critical Periods

For most adults, learning a second language is tough. It often involves countless hours of deliberate effort to assimilate vocabulary, idioms, and grammatical rules before anything resembling fluent processing is achieved in real time. At first blush, the effort typically required to master a second language looks bad for the genetic account of the evolution of language. If inborn linguistic knowledge makes it so easy to acquire a first language, why doesn't it also put wind in the sails of second language learning? However, supporters of the genetic account addressed this problem long ago by arguing that language learning is a "critical period" phenomenon (Lenneberg, 1967; Pinker, 1994). Influenced by research on imprinting and birdsong learning in the 1950s and 1960s, they suggested that the genes make Universal Grammar available to guide language learning only for a limited period in development. The window opens shortly after birth and shuts around puberty. Therefore, children and adults depend on different mechanisms for language acquisition, and, when a language is learned in adulthood, it is rarely mastered to the same extent as when it is learned in childhood.

There are a number of problems with the suggestion that language learning is a critical period phenomenon scheduled by the genes. First, it has become evident since the 1960s that, even in the paradigmatic cases involving imprinting and birdsong, transitions from ease to difficulty of learning are typically experience-dependent rather than genetically programmed (Michel and Tyler, 2005). Second, it is far from clear what the reproductive fitness advantage of switching off Universal Grammar at puberty could be, especially given signs that bi- or multi-lingualism may be typical of human populations (Evans, 2013). Third, there is widespread agreement that the "wild child" literature, suggesting that a first language cannot be acquired properly after puberty, is uninterpretable because it concerns children who have suffered many forms of deprivation and abuse, not merely a lack of linguistic input (Skuse, 1993). Fourth, studies of migrants have shown that proficiency in their new languages correlates, not with whether they arrived before or after puberty, as the critical period view would predict, but with the amount of exposure migrants received to their new languages (Birdsong and Molis, 2001; Flege, Yeni-Komshian, and Liu, 1999; Hakuta, Bialystok, and Wiley, 2003). Finally, recent research suggests that previous studies have overestimated the differences in proficiency achieved by first and second language speakers. Most estimates of first language proficiency are conducted with highly educated people. When first language users with relatively little education are tested, many of them are less able than second language users to comprehend syntactically complex constructions, passives, and sentences using the universal quantifier "every" (Dabrowska, 2012; see below). Second language learners rarely acquire typical phonology—there *are* critical periods for learning to discriminate speech sounds (Werker and Hensch, 2015)—but phonology has very little to do with Universal Grammar. Furthermore, the fact that second language users rarely sound exactly like native speakers could result from an inborn domain-general tendency towards categorical perception (Chang and Merzenich, 2003; Cowie, 2016), and / or from the reuse of phonological representations formed during first language learning (Christiansen and Chater, 2016). It need not be due to genetically inherited language-specific information.

Thus, the critical period debate has not yielded evidence supporting the cultural over the genetic account of the evolution of language. Viewed historically, this debate may look like a defeat for the genetic account. It tried to solve a problem by postulating a critical period for language acquisition, but, for the reasons outlined above, the solution did not work out. This is a reasonable historical interpretation, but it does not tell us now whether the genetic or the cultural account is more likely to be correct. One way or another, they can both explain the relevant data.

RECENT EMPIRICAL DEVELOPMENTS

Linguistic universals and critical periods, discussed in the previous section, have been battlegrounds in the debate about the evolution of language for more than fifty years. This section focuses on three topics—neural localization, sequence learning, and social shaping—which, although not wholly "new," have been investigated with vigor in recent years, often with the intention of providing positive evidence in favor of the cultural account.

Neural Localization

In a meta-analysis of more than 450 functional magnetic resonance imaging (fMRI) studies, Anderson (2008) found that activity during language processing was more widely scattered across the brain than during any other type of task. The spatial distribution of neural activity was greater for language than for reasoning, memory, visual perception, mental imagery, emotion, action, and attention. In a complementary way, Poldrack (2006) found in another meta-analysis of fMRI data, that Broca's area, a region of the brain long famous for being a "language center," was active in more studies involving cognitive tasks unrelated to language (199) than in studies of language processing (166). Furthermore, activation of Broca's area occurred in only 19 percent of all 869 studies in the database that were designed to assess language.

These are substantial and interesting findings. In combination, they suggest that language enlists an unusually wide range of neural structures, and each part of the coalition of brain areas involved in language processing has many other, non-linguistic functions.[2] But the discovery that language, rather than being localized, depends on scattered, multi-functional brain areas does not undermine the genetic account of the evolution of language. It has been repeatedly claimed (Lenneberg, 1967) or implied (Pinker, 1994) that localization of linguistic function, to Broca's area or elsewhere, would support the genetic account over the cultural account, but it is not clear why this was ever thought to be the case (Cowie, 2016). Perhaps it was assumed that genetic selection, but not cultural selection, would favor localization for neural efficiency because long-distance communication among neural networks takes less energy, and is less error-prone, than communication among adjacent networks. If this assumption were valid, the evidence that language is scattered, rather than localized, would count against the genetic account. However, recent "reuse" theories of brain organization suggest that the assumption is not valid. Reuse theories imply that scattering, rather than localization, is the norm. New cognitive functions typically recruit a range of networks, distributed across the brain; networks that had, and continue to have, other functions. Crucially, in the present context, these theories are based on data suggesting that neural recycling is achieved both by genetic evolution (the "massive redeployment" theory; see Anderson, 2010; 2016; Anderson and Finlay, 2014) and, as in the cases of literacy and numeracy, by cultural evolution (the "neural recycling" theory; see

Dehaene, 2009; 2014). So, the dependence of language on scattered, multifunctional brain areas, although incompatible with some long-held assumptions about how genetic evolution organizes the brain, is consistent with both the cultural and the genetic accounts of the evolution of language.

Sequence Learning

The cultural account suggests that language acquisition is powered not by Universal Grammar, but by domain-general processes of sequence learning; in other words, processes that use "statistical" or "associative" principles to encode information about the temporal order of linguistic and nonlinguistic items, presented visually and in the auditory modality. Evidence substantiating this proposal comes from computer simulations, adults, typically developing children, children with "specific language impairment," and non-human animals.

Computer simulations demonstrate that, given a range of linguistic input that is plausibly available to children, domain-general sequence learning networks can acquire the ability to process complex grammatical constructions without built-in features designed by the programmer to handle those constructions. For example, Christiansen and MacDonald (2009) showed that, after exposure to complex recursive constructions, a sequence learning program based on a simple recurrent network (SRN) could process new instances of the recursive constructions, as well as those used in training, and the performance of the network was closely matched with that of human participants.

Other studies, focusing on *human adults,* typically examine "artificial grammar learning." An arbitrary set of rules is used to generate sequences of stimuli (for example, colors, locations, letters, pictograms), and, after a period of exposure to various correct (that is, rule-conforming) sequences, people are asked to judge whether new sequences are correct or incorrect. The advantage of using artificial grammars with adult participants is that the sequences to be learned, although non-linguistic, can be as challenging as the syntactic structures found in natural languages. For example, they can include nonadjacent dependencies, such as "red follows blue" where there are always sequence elements between red and blue. Research of this kind has consistently shown that, in adults, individual differences in artificial grammar learning are positively correlated with syntactic skills, such as processing of subject- and object-relative clause constructions (Misyak and Christiansen, 2012). Furthermore, the findings from these "artificial" studies have been corroborated in naturalistic research involving native English speakers who have encountered the syntactic regularities of Arabic, with minimal semantic cues, through memorization of the Qur'an (Zuhurudeen and Huang, 2016).

Research with *children* has yielded similar results. For example, at six to eight years of age, children's ability to learn the order in which "aliens" enter a space ship predicts their ability to process passives and object-relative clauses, and these predictive relationships hold even when age, verbal working memory, and nonverbal IQ are controlled (Kidd, 2012; Kidd and Arciuli, 2016). Interestingly, the same principle holds for children with "specific language impairment" (SLI). As this term implies, individuals with SLI are commonly understood to have a specific, genetically inherited impairment in processing linguistic grammar (Van Der Lely and Pinker, 2014). However, recent studies with careful controls have shown that, relative to typically developing individuals of the same age, children and adolescents with SLI are impaired in a range of sequence processing tasks (Hsu and Bishop, 2014; Hsu, Tomblin, and Christiansen, 2014; Tomblin, Mainela-Arnold, and Zhang, 2007).

What about *nonhuman animals?* According to the cultural account, the sequence learning processes that support language acquisition are qualitatively similar to those found in other animals. However, in the hominin line, these processes were enhanced by genetic evolution before the emergence of language (Christiansen and Chater, 2016). Consistent with this picture—but also in a straightforward way with the genetic account of the evolution of language—research using artificial grammars with nonhuman animals suggests that, compared with humans, their sequence learning ability is limited. Thus far, there is only evidence that they can encode repetition of sequence elements and the presence of particular units at the beginning or end of a sequence (ten Cate and Okanoya, 2012). However, three developments have provided more specific support for the cultural account. First, a particularly rigorous study found evidence that Old World monkeys are better able than New World monkeys, which are more distantly related to humans, to learn a complex artificial grammar (Wilson et al., 2013). Second, a detailed neurocomputational model has shown how, in principle, syntactic processing could be accomplished by a system that is shared by human and nonhuman primates. In this model, a ventral processing stream is responsible for auditory object recognition, a dorsal stream mediates sequence processing, and language is made possible by better integration of information from the two streams in the prefrontal cortex (Bornkessel-Schlesewsky, Schlesewsky, Small, and Rauschecker, 2015; Ivanova et al., 2016). Third, recent work with mice lends weight to human evidence indicating that FOXP2, rather than being a "language gene," plays a key role in sequence learning more generally (Tomblin, Shriberg, Murray, Patil, and Williams, 2004). Insertion of a humanized version of the gene into mice has a specific effect on cortico-basal ganglia circuits that are important in sequence learning (Reimers-Kipping, Hevers, Paabo, and Enard,

2011) and, at the behavioral level, enhances learning of action sequences (Schreiweis et al., 2014).

Social Shaping

Chomsky's poverty of the stimulus argument assumes that to learn grammar without inborn linguistic knowledge, a child would need not only "positive" input (such as exposure to grammatically correct sentences) but also "negative" input, namely, environmental feedback indicating what is *not* correct, and that children do not receive negative input. The latter assumption was based on evidence that has since been discredited. Chomsky drew on Brown and Hanlon's report (1970) that transcripts of a child, "Eve," interacting with caregivers, indicated that she did not receive negative input. However, Moerk (1991), reanalyzing the transcripts, found many cases in which Eve's semantic and syntactic errors were corrected by her interlocutors. Nonetheless, the idea has lingered, among supporters of both the genetic and cultural accounts of the evolution of language (Christiansen and Chater, 2016), that there is a paucity of negative input for language learning.

This lingering doubt is curious because there is, and has been for some time, a substantial body of evidence showing that plenty of negative input is available (Bohannon, MacWhinney, and Snow, 1990; Cowie, 2016). For example, analyzing transcripts from two- to three-year-old children interacting with their parents and other adults, Bohannon and Stanowicz (1988) found that about one-third of all phonological and syntactic errors received negative feedback—they were rejected, or repeated with correction—and, in common with Demetras, Post, and Snow (1986), that incorrect sentences are much less likely than correct sentences to be repeated verbatim by parents. Although it is not clear why this transcript-based evidence has been largely ignored by both parties to the debate on the evolution of language, two recent developments may lend it greater prominence. The first is a trend towards observational and experimental studies demonstrating that negative input is not merely available but has a significant impact on language learning. For example, in a longitudinal study of mothers and their children, Taumoepeau (2016) found evidence that a mother's spontaneous tendency to "expand" her child's utterances—to repeat the meaning while supplying missing syntactic information—contributed to vocabulary learning.

Leading the second development, Dabrowska (2012) has found that native English speakers with less than eleven years of formal education, and those with at least seventeen years of formal education, do not have the same grammatical knowledge. For example, less educated speakers have difficulty comprehending possessive locative sentences, which cannot be explained by lack of motivation, or stress in the experimental environment. When given a sentence such as "Every fish has a bowl," and asked to say which of two pictures the sentence describes, less educated speakers are as likely to choose a picture of three bowls, each containing a fish, plus one fish without a bowl, as they are to choose a picture of three bowls, each containing a fish, plus one bowl without a fish (Street and Dabrowska, 2010, Experiment 1). If there is diversity of grammatical knowledge among native speakers, it need not be due to variation in the negative input they have received. However, confirmation that such diversity exists would put a premium on discovering its sources, and a role for negative input is suggested by a follow-up study in which explicit training, including correction, improved comprehension of possessive locatives by participants with relatively little education (Street and Dabrowska, 2010, Experiment 2).

BROAD ARGUMENTS

The research reviewed in the previous section—on neural localization, sequence learning, and social shaping—seems, like work on universals and critical periods, to support the cultural over the genetic account of the evolution of language. It confirms novel predictions of the cultural account and in several cases—relating to neural localization, "specific language impairment," FOXP2, and negative input—undermines findings that were previously regarded as evidence in favor of the genetic account. However, despite what appears to be a long series of empirical defeats, the genetic account has not fallen. Supporters continue to argue, plausibly, that the new data: (1) are off target (for example, linguistic universals are not features common to all natural languages); (2) relate to inessential features of their theory (for example, there must be language genes, but FOXP2 need not be one of them); or (3) can be embraced by their theory without concession. For example, Lidz and Gagliardi (2015) suggest that all of the recent data indicating the power of domain-general processes of sequence learning are entirely compatible with the existence of Universal Grammar. Sequence learning merely provides the "inferential" component of the language learning system, while Universal Grammar provides the "deductive" component.

As an outsider looking over the fence, I find the resilience of the genetic account very puzzling. It seems that any other theory in cognitive science that had received so many empirical knocks would have lost credibility long ago. The fact that the genetic account continues to retain and recruit supporters raises two related questions: Why is it so resilient, and what, if anything, would count against it?

Three factors appear to have contributed to the resilience of the genetic theory. First, it is a heterogeneous, moving target for empirical inquiry. All contemporary versions of the genetic theory are rooted in Chomsky's work on Universal Grammar, but some claim that language emerged gradually, through a conventional process of adaptation, whereas others suggest it appeared suddenly, as a result of chance mutation. Furthermore, claims about the content of Universal Grammar have changed radically since the 1950s—from transformations, to principles and parameters, to the minimal program—and each of these views continues to be represented in contemporary linguistics. Divergence of opinion, and theory change, are natural, healthy parts of scientific inquiry. However, the extent of the diversity and mobility of the genetic account of language make it very difficult to test. A result that counts against one version may well be compatible with another. Second, the isolation of competence from performance contributes to the resilience of the genetic theory. This distinction, which has long been at the heart of Chomsky's work, is now cast as a divide between a genetically inherited "language of thought" (competence) and processes of "externalization" (performance), which, by various means, convert an instrument of thought, Universal Grammar, into an instrument of communication, natural language (Berwick and Chomsky, 2015). This is, of course, an entirely coherent set of claims about the way the mind works. However, like previous incarnations of the competence-performance distinction, it threatens to insulate hypotheses about Universal Grammar from evaluation by cognitive science. Anything the cognitive scientist can find out about natural language acquisition and processing—in studies of infants, adults, and nonhuman animals; through naturalistic studies, experiments, and computer simulations—is liable to be regarded as a discovery about externalization, with no bearing on the nature or origins of Universal Grammar.

Finally, and more speculatively, the genetic theory may be resilient because it engages historically deep convictions about language and its role in defining human nature (Eco, 1995). The idea that language is in our genes, and only *our* genes, draws a bright line, a Rubicon, between humans and other animals. It also suggests that something precious to us—the capacity to express and understand each other's thoughts—is secure. A culturally constructed capacity could be lost or eroded by catastrophe, but a genetically inherited capacity is safely locked inside, reemerging, fresh and complete, in each new generation. We who look over the fence may even hope that the linguistic knowledge locked in our genes is a Platonic knowledge of objective, eternal realities. Although there is a Platonic version of Chomsky's theory (Katz, 1985), it cannot be combined coherently with an evolutionary perspective. The combination would imply that genetic selection, a natural process, has produced a set of adaptations—Universal Grammar—to an *un*natural environment of ideal forms. Given this incoherence, it is very unlikely that any expert on language would embrace an evolutionary-Platonic view, but the resilience of the genetic theory does not depend only on expert opinion. Unlike many topics in cognitive science, language is important to "outsiders" as well as "insiders." Our deeply held, but not always rational, beliefs about language could help to sustain the genetic theory in the face of apparently contrary evidence, or signs that it is becoming immune to empirical testing.

Supporters of the cultural account have tried to make two very broad arguments count against the genetic theory. The first raises what Christiansen and Chater (2016) call "the logical problem of language evolution," referencing Chomsky's "logical problem of language acquisition." For example, they argue, informally and via computer simulations (Chater, Reali, and Christiansen, 2009), that Universal Grammar could not have evolved genetically because linguistic conventions change too quickly (Gray and Atkinson, 2003). As long as a convention, A, is stable, there could be selection pressure in favor of genes that facilitate learning of A, that is, genetic assimilation or a "Baldwin effect" (Pinker and Bloom, 1990). However, as convention A gives way to convention B—due to innovations within the speech community or external forces such as language contact—there will be selection against the A-learning genes, and the beginnings of a genetically evolved Universal Grammar will be swept away. It would be circular to suggest that this problem does not arise because Universal Grammar stabilizes linguistic conventions. The question is whether Universal Grammar could become genetically fixed, and, therefore, the answer cannot assume that Universal Grammar has already been fixed. Neither, according to Christiansen and Chater, can the problem be solved by genetic drift. Intuitively, the following scenario is plausible: random genetic variation in a population makes it easier for members to learn convention A than to learn convention B; this "genetic preference" results in more use of A than of B in the linguistic environment; and this bias in the linguistic environment creates selection pressure favoring genes for A over genes for B. Although intuitively plausible, modeling suggests that this kind of effect would occur only if the genetic preference favoring A over B was very strong at the outset (Chater et al., 2009). Random variation with a strong effect is less likely than random variation with a weak effect, and an initially strong genetic preference would leave little work to be done by genetic assimilation.

The foregoing is a sketch of just one part of Christiansen and Chater's logical problem of language evolution (2016; Chater and Christiansen, 2010). In its entirety, the argument is broader, deeper, and supported by extensive modeling. Perhaps, as it becomes better known, it will weigh heavily against the genetic account, but my hunch is that an argument about what could or could not have happened in the distant past is unlikely to prove decisive. If the genetic account resists damage by relatively unequivocal data, collected in the here-and-now, it is unlikely to be felled by inferences about rates of linguistic change in ancestral environments.

The other kind of broad argument made against the genetic theory is, at its core, a parsimony argument (Christiansen and Chater,

2016; Culicover, 1999; Tomasello, 1995). It suggests that data of the sort reviewed in this chapter make Universal Grammar unnecessary. Contrary to the poverty of the stimulus argument, data show that domain-general processes are sufficient for language learning. Therefore, theories postulating roles for both domain-general processes and Universal Grammar (Lidz and Gagliardi, 2015), even if they can be made consistent with the data, are engaging in unnecessary duplication. A complementary line of argument is not open to supporters of the genetic account. The distinction between competence and performance—between Universal Grammar and processes responsible for learning a particular natural language—is so deeply embedded in the genetic account that its supporters cannot now argue that Universal Grammar is potentially sufficient for language learning.

Perhaps the parsimony argument will ultimately prove to be decisive, but it has stiff opposition. We may not need Universal Grammar, but many of us want it. It is wanted by linguists to define their professional boundaries, by evolutionary psychologists as a paradigmatic example of a cognitive instinct, and possibly, at some level, by all of us as a bright line separating humanity from the rest of the animal kingdom.

CONCLUSION

This chapter has outlined a genetic and a cultural account of the evolution of language, and reviewed evidence relating to linguistic universals, a critical period for language development, the neural localization of language, and the roles of domain-general sequence learning and social shaping in language acquisition. The discussion of linguistic universals indicated that there are few, if any, non-definitional features that all languages have in common. However, this can be made compatible with the genetic account if linguistic universals are construed not as features that all (or many) languages have in common, but as components of Universal Grammar. The discussion of critical periods drew attention to evidence from migrant populations suggesting that second language proficiency depends on number of years of exposure to the second language, rather than on whether learning began before or after puberty, and to studies of native speakers indicating that, with the exception of phonology, first and second language learners may obtain similar levels of proficiency. These findings suggest that language learning is not a critical period phenomenon, but the critical period claim is not an original or essential part of the genetic account of the evolution of language.

When looking at the neural localization of language, we found that it enlists a more widely distributed set of brain areas than any other major psychological function, and that Broca's area is more often active during non-linguistic than linguistic tasks. These data certainly tell against the idea that there is a "language center," but it is not clear why it was ever supposed that genetically inherited linguistic information is more likely than culturally inherited information to be implemented in a narrowly localized area of the brain.

The review of research on domain-general sequence learning highlighted several sources of evidence that are consistent with the cultural account of the evolution of language. Computer simulations suggest that sequence learning, without inbuilt language-specific constraints, can enable a system to process complex grammatical constructions in a humanlike way. Experiments examining individual differences in typically developing adults and children suggest that they use the same sequence learning processes to learn artificial and "real" linguistic grammars, and studies of people with "specific language impairment" indicate that their impairment is not, in fact, specific to language; they have difficulty with sequence learning across task domains. Likewise, research with nonhuman animals confirms that domain-general sequence learning capacity has increased in the hominin line, provides a plausible model of how this change has been implemented in the primate brain, and supports evidence from humans that mutations of FOXP2 interfere with language by interfering with sequence learning more generally, implying that FOXP2 is not a "language gene." Finally, research on social shaping shows that infants and children are frequently corrected by adults when they make grammatical errors, and that this negative input is put to use in language learning. These findings, like those on sequence learning, confirm novel predictions of the cultural theory and, in the case of social shaping, challenge the poverty of the stimulus argument, a foundation of the genetic account. However, those who are committed to a firm distinction between competence and performance can argue that all of these findings bear on the externalization of language, but not on whether there is a genetically inherited language of thought.

When I started researching this chapter, I was ready to be convinced that language is a cognitive instinct rather than a cognitive gadget. I was prepared to accept that, while other distinctively human cognitive processes are products of cultural evolution, language really is in our genes. Indeed, it would make my job easier to have one, foundational instinct on which to build. But, for what it is worth, this outsider has not been convinced. Maybe there really is a genetically inherited Universal Grammar, but, from over the fence, it looks like that theory cannot now be tested against the cultural evolutionary alternative using the methods of cognitive science. I can only conclude that, while the genetic view is appealing for a variety of reasons, some of them extra-scientific, the cultural account—once a very poor relation—is now clearly specified and rich in empirical support.

9

CULTURAL EVOLUTIONARY PSYCHOLOGY

WE HAVE LOOKED IN DETAIL AT FOUR DISTINCtively human cognitive mechanisms: selective social learning, imitation, mindreading, and language. In the first three case studies, I argued that the evidence from cognitive science—encompassing research from experimental psychology and cognitive neuroscience, as well as comparative, developmental, and social psychology—indicates that selective social learning, imitation, and mindreading are cognitive gadgets rather than cognitive instincts. In the case of language, I am not qualified to judge the balance of evidence. However, it is clear, even to an outsider, that the genetic view no longer occupies the unassailable position it once enjoyed.

With the case studies under our belt, it is now time to step back and reconsider broader issues relating to the cognitive gadgets theory, and to examine the implications of this framework for research on human cognition—the prospects for a cultural evolutionary psychology. The first three sections of this final chapter tackle evolutionary issues introduced in Chapter 2: What are the selection processes and inheritance mechanisms involved in the cultural evolution of cognitive processes? Are they likely to have been genetically assimilated? The subsequent three sections step back yet further to consider how the cognitive gadgets theory can be related to human evolutionary history, what it implies about human nature, and how it can be used in psychology and other disciplines to investigate the origins and functions of distinctively human cognition.

CULTURAL GROUP SELECTION

In Chapter 2, I distinguished selectionist models of cultural evolution from those that are merely populational. In all populational models, the frequency distribution of types within a population at time T+n is largely determined by their distribution at time T. In the subset of populational models that are selectionist, the frequency of types at time T+n is determined in a specific way by the distribution at time T: by mechanisms of variation, sorting, and heritability ensuring that types which are beneficial at one time step increase in frequency at the next *because* they are beneficial (Sterelny, in press).

Some cultural evolutionists, focused on grist (behavior, artifacts) rather than mills (cognitive mechanisms), say that selection is an important force, but they do not address squarely the key questions that a selectionist view entails. Specifically, some members of the California school suggest that the fitness of cultural traits, the extent to which they proliferate, depends on the benefits they confer, but they do not state plainly, or in a consistent way: (1) who benefits—the individuals or groups that bear the traits (for example, experts, thinkers) or the traits themselves (for example, skills, beliefs); and (2) the nature of the benefit—in what sense the design of the trait influences the fitness of the beneficiary (Clarke and Heyes, 2017; Dennett, 2001). This has led to uncertainty about whether the California school is really committed to a selectionist view of cultural evolution, and, if not, whether or in what sense their models explain the adaptiveness, the well-designedness, of cultural traits (Lewens, 2015; Sterelny, in press.)

To avoid similar uncertainty, I want to make as clear as possible what I have in mind when I propose that distinctively human cognitive mechanisms are products of cultural evolution. I am suggesting that they are shaped by cultural group selection.

Imagine a population divided into two social groups, "left" and "right." In this context, a social group is a group of people defined not by the genes they carry, but by geography and / or cultural characteristics such as language. People in the left group are more likely to have one version of a cognitive mechanism, M, and people in the right group are more likely to have an alternative version, M′. The people in both groups have biological offspring, and these biological offspring culturally inherit—that is, develop through social learning from their parents and other members of the group—their own version of the focal cognitive mechanism. Thus, children born into the group where people tend to have M are more likely to develop M, while children born into the group where people tend to have M′ are more likely to develop M′.

Now suppose that M′ is better than M in fulfilling some function (copying body movements, for example) and, therefore, bearers of M′ tend to have greater success in meeting some goal (for example, obtaining food). Consequently, people with M′ have more babies than people with M. The M′ group also attracts recruits from the M group—people who observe the success of the other group, choose to migrate into it, and replace their M mechanism with the superior M′. The M′ group is thus fitter than the M group, where the fitness of a social group can be understood in either of two ways. The M′ group might be fitter because it acquires a greater proportion of the total population

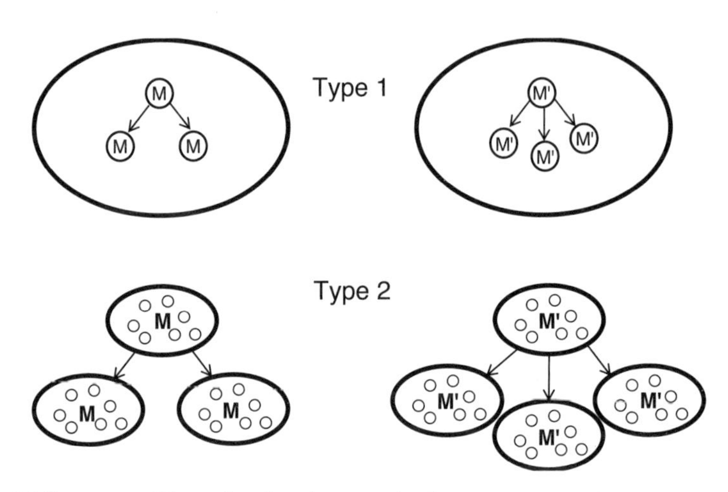

9.1 Two types of fitness in cultural group selection.

as its members (Type 1 fitness), or because it founds a greater number of descendant groups—groups which inherit the use of the M′ mechanism (Type 2 fitness).

The distinction between the two types of fitness is derived from the distinction between two kinds of "multilevel selection," known as MLS1 and MLS2 (Damuth and Heisler, 1988; Okasha, 2005). Figure 9.1 represents the two possibilities, with the pair of ovals at the top representing cultural group selection with Type 1 fitness, and the set of ovals at the bottom representing cultural group selection with Type 2 fitness. In the Type 1 case, individuals (circles) with M′ typically have three biological offspring with M′, whereas individuals with M have just two. In the Type 2 case, M′ groups—in which there are more individuals with M′ than M—typically bud, or undergo fission, to produce three descendant M′ groups, whereas M groups—in which there are more individuals with M than M′—typically produce two M descendant groups. Rates of biological reproduction and migration may influence rates of production of descendant groups. For example, larger groups may be more likely than smaller groups to bud. Either way, there is selection in favor of M′ in the sense that, over generations, the number of people with M′ increases relative to the number of people with M. Note, however, that the selection is cultural rather than genetic, not only because the cognitive mechanisms M and M′ are inherited through social learning, but also because social group membership influences which cognitive mechanism an individual is likely to develop.

Who benefits? When M′ increases relative to M, we can say that mechanism M′ itself benefits, insofar as it is the winner of the competition. Since M′ helps individual people to achieve some goal, its human bearers also benefit. They meet their goal and have more babies. But there is a sense in which the social group benefits, too, insofar as it is good for the group to have a greater number of members, or a greater number of descendant groups. Anthropological evidence is needed to distinguish these three possibilities; to work out who, in fact, typically benefits from cultural group selection of cognitive mechanisms.

What is the nature of the benefit? Cognitive mechanisms modulate the fitness of their bearers via their effects on living conditions, including food, shelter, and defense against predators. Individuals with better living conditions are likely to have more children that survive and reproduce (Type 1 fitness), and groups with better living conditions are more likely to persist through time and to bud, not only because their members are more likely to survive and reproduce, but also because these groups are more likely to attract net immigration (Type 2 fitness). Groups with better living conditions are also more likely to have their practices emulated by other groups, including child-rearing and ritual practices that foster the development of particular cognitive mechanisms. Therefore, enhanced living conditions may also foster group fitness (Type 2), and increase the number of descendant groups, by encouraging "group copying": copying of one group by other groups.

Asocial and social cognitive mechanisms deliver better living conditions in different ways. Asocial cognitive mechanisms promote the discovery of strategies and technologies to harness and defend natural resources (for example, causal understanding). Social cognitive mechanisms both enable these strategies and technologies to be learned by others, thereby promoting the cultural evolution of grist (the social cognitive mechanisms that do this are called cultural learning), and enable cooperation among group members. For example, configural face processing, a social cognitive process that is not a form of cultural learning, helps group members keep track of who is and who is not pulling her weight, and police cooperation by rewarding labor for the group and punishing free-riding. Some social cognitive mechanisms, including imitation, mindreading, and language, benefit the group by promoting both cooperation and the cultural inheritance of grist.

Let's take imitation as an example. There is one imitation mechanism inside each person's head, which gears motor sequence learning to perceptual sequence learning via matching vertical associations (ASL model, Chapter 6). The M′ version of this mechanism has a richer repertoire of matching vertical associations for whole body movements than the M version, enabling people with M′ more accurately to imitate actions involved in ritual (for example, dance), hunting (for example, stalking), and combat (for example, spear throwing). As a consequence, bearers of M′ are better able than bearers of M to cooperate in a range of tasks—accurate imitation of ritual movements promotes synchrony, and synchrony promotes social bonding (Heyes, 2013; Tarr et al., 2015; Tunçgenç and Cohen, 2016)—and to sustain the cultural inheritance of techniques that enhance success in hunting and intergroup combat. These advantages may lead groups in which the M′ mechanism predominates to acquire greater numbers of new members, or to produce more descendant groups, than groups in which M predominates.

INHERITANCE

All selection processes require robust inheritance mechanisms: processes that preserve and propagate variants (Chapter 2). In the case of genetic evolution, robust inheritance is achieved primarily by DNA replication. For the cultural evolution of grist—knowledge and skills—it is achieved by social and cultural learning.[1] What are the inheritance mechanisms for the cultural evolution of mills—of cognitive mechanisms, including the mechanisms of cultural learning?

At the cognitive level—in terms of what goes on inside individual heads—the inheritance mechanisms for mills overlap with the inheritance mechanisms for grist. For example, explicit metacognition, something very like mindreading, is important in the development of genuinely strategic social learning (Chapter 5), and language, the production and comprehension of sign sequences, plays a crucial role in the acquisition of mindreading (Chapter 6). However, the inheritance of cognitive mills can also involve group-level or social processes—things that go on between people, rather than inside individuals' heads—such as conversation, storytelling, turn-taking, collective reminiscing, teaching, demonstrating, and engaging in synchronous drills. All of these social processes, and no doubt many more, shape the development of cognitive mechanisms. Through conversation, teaching, and demonstration, children learn to deploy metacognitive social learning strategies in the same way as the people around them (Chapter 5). Through turn-taking in face-to-face interaction, and engaging in synchronous drills, children acquire a particular repertoire of matching vertical associations; they become able to imitate the same range of actions as their cultural parents (Chapter 6). Through conversation, storytelling, and collective reminiscing, children become able to represent mental states and accumulate a stock of generalizations about the way mental states relate to one another, to behavior, and to the world (Chapter 7; Salmon and Reese, 2016; Nile and Van Bergen, 2015).

Dedicated research of a radically new kind is needed to measure the robustness of these inheritance mechanisms, the fidelity with which distinctively human cognitive mechanisms are passed down via social processes from one cultural generation to the next. In advance of such research, three considerations suggest that they are robust enough to support cultural group selection of cognitive processes. First, selection can act as long as, and to the extent that, the characteristics of parents and their offspring are correlated. High fidelity replication, of the kind supported by DNA, is not a necessary condition for Darwinian selection (Godfrey-Smith, 2012). Second, distributed inheritance, which is likely to be typical of social cognitive mechanisms at least (Chapter 2), has a better chance of being robust than inheritance that occurs via one route—vertical, oblique, or horizontal. When inheritance is distributed, if one route is blocked—for example by death of a parent (vertical), childhood illness (oblique), or introversion (horizontal)—the other routes can compensate. Third, the case studies examined in Chapters 5–8 suggest that a range of social processes can contribute to the cultural inheritance of each cognitive mechanism, providing another source of redundancy. The imitation mechanism is constructed through imitation of children by adults, exposure to optical mirrors, engagement in synchronous drills, *and* acquired equivalence experience (Heyes and Ray, 2000). Similarly, the mindreading mechanism is built by conversation, storytelling, *and* collective reminiscence. Finally, each of the social processes occurs repetitively. Children are told a particular story not once but many times; different stories contain the same themes, morals, and tropes; adults imitate the same facial gestures over and over again in face-to-face interaction with infants; collective reminiscence returns repeatedly to the same episodes. Thus, children receive many learning trials. They have many opportunities to pick up and consolidate the same information.

The case of print reading shows clearly that social processes can support robust cultural inheritance of cognitive mechanisms when they are designed to do so (see Chapters 1 and 7). In literate societies, cognitive mechanisms dedicated to reading are reliably reconstructed in each successive generation through epistemic engineering—for example, giving children picture books with easy-to-read words—and formal instruction by parents and teachers. And this does not occur by chance, or as a consequence of a blind selection process. Literacy training is *intended* to make children literate; it has been designed by educators to have exactly that effect. In contrast, as far as I am aware, in no cultures are all children deliberately trained to use metacognitive social learning strategies, to imitate, interpret the actions of others, or use language. Consequently, we cannot infer from the success of literacy training that other cognitive mechanisms are faithfully inherited. However, this disanalogy between literacy and other distinctively human cognitive mechanisms does have a couple of interesting implications.

First, the disanalogy implies that, at present, many other distinctively human cognitive mechanisms, beyond literacy and numeracy, owe their adaptive characteristics to selection, not "design." Both grist and mills are "adaptive," or have "adaptive characteristics," if they do their jobs well. A fishing technique (grist) is adaptive if it results in a big catch; a mindreading mechanism (mill) is adaptive to the extent that it allows accurate prediction of behavior. In principle, both grist and mills could be adaptive because they have been shaped by:

(1) genetic selection, (2) "intelligent design," or (3) cultural selection (Heyes, in press a). In this context, intelligent design would involve one or more humans using "practical intelligence" (Godfrey-Smith, 2012), or causal understanding, to devise an efficient way to catch fish, or to promote the development in children of an efficient mindreading mechanism. In the case of grist, it is often very hard to work out whether the adaptive characteristics of a trait are due to intelligent design or cultural selection, items (2) and (3), above. A particular fishing technique *could* be the result of a selection process, in which ineffective elements have been winnowed out and effective elements preserved, without anyone thinking hard about why some parts are effective and others not. But for most fishing techniques, where little is known about their history, and it would be easy for people to compare their effectiveness in number of fish caught, it is at least equally plausible that the technique was devised by intelligent design—in one shot of insight or via successive rounds of deliberate experimentation. This problem does not haunt the cultural evolution of mills. With a few exceptions in contemporary industrialized societies, including literacy and numeracy, we can be confident that the adaptive characteristics of cognition mechanisms are not due to intelligent design. For example, it is highly implausible that any smart person, or design council, ever anticipated that self-observation, being imitated, exposure to mirrors, and / or performing synchronous drills would be good for the development of effective imitation mechanisms.

The disanalogy between literacy and other distinctively human cognitive mechanisms also implies that, in the future, the cultural inheritance of other cognitive mechanisms could be enhanced by formal education. It may be possible to design training programs—for use by caregivers, in schools, or in the criminal justice system—to improve cognitive skills such as selective social learning, imitation, and mindreading in a whole society or particular group. In these cases, as with literacy and numeracy, cultural selection could be augmented by intelligent design.

GENETIC ASSIMILATION

Many cognitive mechanisms, like imitation and mindreading, not only do their jobs well, but do jobs that, when done well, seem likely to enhance reproductive fitness—to increase the number of babies produced by the bearers of the cognitive mechanisms. This has led some researchers to assume that, even if new cognitive mechanisms are produced by learning in a culture-soaked environment, they will later become genetically assimilated. In other words, they may start out as cognitive gadgets, constructed in the course of development through social interaction, but then selection will progressively favor genetic mutations that reduce the experience-dependence of the gadgets' development, converting them into cognitive instincts (Henrich, 2015).

As I stated in Chapter 4, and justified in Chapters 5–8, I believe the genetic assimilation hypothesis is at odds with the evidence from cognitive science. Time and again the evidence indicates wealth, not poverty, of the stimulus: covariation between the development of distinctively human cognitive mechanisms and opportunities for learning (Chapter 2). This covariation does not rule out, in principle, the possibility that genetic evolution has speeded up the relevant learning processes. However, I have not been able to find positive evidence that this kind of genetic assimilation has occurred—for example, evidence that learning is faster in natural than unnatural conditions, or that identical twins are more alike than fraternal twins. Indeed, in cases where positive evidence of genetic influence has been sought, the signs have pointed in the opposite direction. For example, people are not slower to associate body movements with unnatural stimuli, events that our ancestors would not have encountered, and identical twins are no more alike in their imitative ability than fraternal twins (McEwen et al., 2007; see Chapter 6). So, the current evidence suggests that our cognitive gadgets have not been genetically assimilated. But if this is true, *why* is it true?

There are a number of potential answers. It could be that cognitive gadgets have not been genetically assimilated because they are locally but not globally optimal, or that genetic assimilation has been obstructed by fitness valleys, or by lack of appropriate genetic variance (West-Eberhard, 2003; 2005). But my guess is that the most important factor is the speed of environmental change. Distinctively human cognitive mechanisms need to be nimble, capable of changing faster than genetic evolution allows, because their job is to track specific, labile features of the environment. For example, social learning strategies track "who knows" in a particular social group, something that changes with shifting patterns in the division of labor and, therefore, of expertise. Imitation tracks communicative gestures, ritual movements, and manual skills that change as groups find, through the cultural evolution of grist, new group markers, bonding rituals, and technologies. And mindreading, like language, must not only track externally driven change in the phenomena it seeks to describe—for example, economically and politically driven fluctuations in the degree to which behavior really is controlled by social roles and situations rather than beliefs and desires—but also self-generated change. Because it has regulative as well as predictive functions (McGeer, 2007), changes in mindreading can alter their explanatory target—the way the mind actually works.[2]

In short, distinctively human cognitive mechanisms are tracking targets that move too fast for genetic evolution. In a stable phase, "assimilative alleles"—genes that reduce the experience-dependence of a cognitive gadget's development—may increase in frequency. But when the environment shifts, there will be selection against assimilative alleles because their bearers will be slower to adjust to the new conditions (Chater et al., 2009). Once again, let's take imitation as an example. As long as gestural markers of group membership, bonding rituals, and technologies remain constant, alleles that privilege and accelerate learning of particular matching vertical associations could be targets of positive selection. For example, people who more readily associate matching trunk movements (for example, you lean forward, I lean forward) than complementary trunk movements (you lean forward, I lean back), might have higher reproductive fitness than people who learn matching and complementary trunk movements at the same rate. But when conventions or technologies change, those assimilative alleles would hamper the development of imitation mechanisms with a now more effective repertoire of matching vertical associations. The people who had once been such effective social operators would now be losing social capital by leaning in when they should be leaning back. This kind of problem could be avoided if mutation produced a universal imitation mechanism, like the cognitive instinct postulated by Meltzoff and Moore (1997), which could copy the topography of any body movement. However, this would be standard genetic evolution, not genetic assimilation, and, given that no one has worked out how such a mechanism could operate (Chapter 6), it is plausible that—like wheels (Dennett, 1984)—it lies outside the range of available genetic variation.[3]

In contrast with cognitive gadgets, the components of the starter kit (Chapter 3) are ripe for genetic assimilation because they do nonspecific jobs that continue to be worth doing in spite of rapid and radical change in human social environments. Social tolerance and motivation promote the development of cooperation whether people are shifting rocks or designing rockets together. Attending closely to faces and voices opens a floodgate of information from other people, whether the information is about the value of a root or a roux, and high power associative learning and executive function improve problem-solving across a huge range of social and asocial problems. Changes to cognitive mechanisms that increase the supply of information from social sources, and the efficiency of problem-solving across domains, are good targets for genetic assimilation because they remain adaptive as long as the developmental environment contains knowledgeable agents and tricky problems to be solved. But changes to cognitive mechanisms that tune human development to specific features of the culture-soaked environment—cognitive gadgets—are poor targets for genetic assimilation because they remain adaptive only until those features change.

A LITTLE HISTORY

Now we turn from the cognitive gadgets theory itself—what it says about cultural group selection, inheritance mechanisms, and genetic assimilation—to the broader implications of the theory. In Chapter 1, I introduced the distinction between narrative and force theories of human evolution. The cognitive gadgets hypothesis is a force theory; it is concerned with the processes involved, rather than the history of events, in human evolution. The ideal theory would be synthetically high on both the historical and force dimensions—it would use chronology as evidence of forces, and forces to explain chronology. Therefore, connecting the cognitive gadgets theory to key events in human evolution, using the archaeological record, is a priority for future research. In this section, I make a start in that direction. I am not an archaeologist, so I use a broad brush, but I hope to give a sense of how the cognitive gadgets theory could be used by specialists to interpret the archaeological record.

Simple stone tools began to be used by hominins about 3.2 million years ago (MYA). More complex Acheulian tools appeared about 1.7

MYA, and there are early signs of hafted and blade-like tools about 250 thousand years ago (KYA). Thus, the archaeological record suggests that hominin technology—and, by implication, ways of living—changed relatively little over a three million year period. Then something big happened. Until recently, it was the received view that Homo sapiens underwent an abrupt upgrade in cognitive sophistication—becoming "behaviorally modern"—between fifty and forty KYA, and that this "Upper Palaeolithic Revolution" was driven by genetic change (Mellars, 1989; 2005; Mellars and Stringer, 1989). In the last twenty years, this view has been undermined by signs that the light of cognitive complexity was not suddenly switched on in one place at one time (McBrearty and Brooks, 2000). Instead, there was "flickering." Over a 100–200 KY period to forty KYA, various signs of cognitive sophistication—for example, evidence of an expanded trade network, bow and arrow technology, jewelry, and ornaments—appeared and then disappeared in different regions of Africa. This new information makes the transition to behavioral modernity look less like a revolution, but it is still a major explanatory challenge (Sterelny, in press).

Dual inheritance theorists have formulated an influential response to this challenge known as the "distributed cognition" or "collective intelligence" hypothesis (Henrich, 2004; 2015; Muthukrishna and Henrich, 2016; Richerson and Boyd, 2013). This hypothesis suggests that the emergence of behavioral modernity was due to climate-driven demographic changes that enabled gene-culture coevolution to get off the ground. Previously, individual hominins—at least those descending from *Homo heidelbergensis*—had, as part of their genetic endowment, the psychological capabilities necessary for cultural evolution, but adaptive innovations of grist—knowledge and skills—did not emerge, persist over generations, or get improved because social groups were too small and weakly connected with one another. These earlier hominins met the endogenous requirements for cultural evolution, they were smart enough, but not the exogenous requirements; not enough of them were in regular contact with one another. In small, weakly connected groups, adaptive innovations are less likely to be generated and passed on. Adaptive innovations are less likely to be generated because the economics of a small group typically prevent specialization. They do not allow some individuals to become the kind of expert—for example, in tool-making or the use of medicinal herbs—who is likely come up with useful new techniques. Any new techniques that do emerge are less likely to be passed on because, when only a small number of people know something—for example, how to tie a particular knot—there are few potential models from which to learn. Consequently, lack of opportunity for novices to observe the skill, due to death or inaccessibility of the models, is likely to result in the skill's extinction (Sterelny, in press). However, according to the collective intelligence hypothesis, around 250 KYA, humans started living in larger and better connected foraging bands, and cultural evolution began to work its magic, albeit in a faltering, flickering way, due to regional and global disturbances that sometimes forced social groups to disperse.

Some modeling and ethnographic evidence make it plausible that, as the collective intelligence hypothesis proposes, the transition to behavioral modernity was driven by demographic change launching cultural evolution. For example, there is a positive correlation between population size and technical complexity across the islands in Oceania (Kline and Boyd, 2010). However, the evidence from cognitive science, reviewed in this book, suggests a different picture of what the demographic changes did to human minds. The collective intelligence hypothesis implies that they did not change how human minds work, only what those minds contained. As groups got larger, minds became packed with culturally inherited grist—knowledge and skills. In contrast, my analysis, the view from cultural evolutionary psychology, suggests that the demographic changes also launched the cultural evolution of mills—of Big Special, distinctively human cognitive mechanisms. The Small Ordinary components of the genetic starter kit (Chapter 3) were already in place and had been supporting cooperation and simple stone technologies for millions of years. Demographic changes allowed the Small Ordinary components to begin to be elaborated by cultural group selection into the mechanisms that we now identify as, for example, causal understanding, episodic memory, imitation, theory of mind, and full-blown language. As archaeologists have long suspected, the emergence of behavioral modernity involved a major cognitive upgrade. However, the primary locus of change was in *how* people thought, rather than *what* they thought, and the changes were wrought primarily by cultural rather than genetic evolution.

HUMAN NATURE

What is human nature? Many philosophers of science now believe this is a bad question, that the concept of human nature is flawed and anachronistic (Lewens and Hannon, in press; Hull, 1986; Lewens, 2012). If one believes that humans are fundamentally different from everything else in the natural world, perhaps specially created by God, it makes sense to look for sharp boundaries—necessary and sufficient conditions for being human. But, the critics argue, we now know that humans are products of evolution, not a natural kind, but a segment of a lineage of creatures that worms its way through time. We have distinctive observable features—for example, we are naked apes—but we do not have internal properties, present in us and absent in all other animals, which make us who we are. We do not have essences or souls. Other researchers, including myself, think there is still some value in the concept of human nature, that it is helpful in circumscribing scientific projects—for example, the purposes of evolutionary psychology—and as a way for philosophers and scientists to communicate, in academia and in public life, what their work implies about the distinctive features, origins, and flexibility of human behavior.

Two recently formulated theories, Machery's "nomological account" (2008; in press) and Samuels's "causal essentialist" theory (2012), are designed to help the concept of human nature fulfil these functions. The nomological account says that "human nature is the set of properties that humans tend to possess as a result of the evolution of their species." It combines the "the universality proposal," that traits belonging to human nature must be typical of human beings, and "the evolution proposal," that they must be products of genetic evolution (Machery, in press). In contrast, the causal essentialist view says that "human nature is a suite of mechanisms that underlie the manifestation of species-typical cognitive and behavioral regularities" (Samuels, 2012).

Both of these theories escape critics' primary objection to the concept of human nature by linking it with characteristics that are typical of humans, rather than rigidly necessary and sufficient to be human. The distinctive strength of the nomological account is that it puts evolution in the foreground and is, therefore, immediately compatible with many projects in the human sciences, including sociobiology and human behavioral ecology as well as High Church evolutionary psychology. But the nomological account also has two weaknesses. First, it allows all traits we possess as a result of the evolution of our species—morphological, physiological, behavioral, and cognitive traits—to be part of human nature. Consequently, the nomological view threatens to make human nature into a long, dull list of features that could be used by a Martian as a "field guide" to identify humans among other earthly creatures, but does not distinguish important characteristics from trivial ones (Lewens, 2015). Second, the nomological account seems to take a narrow and outdated view of evolution. It embraces human characteristics that have been shaped by selection acting on DNA sequences, but not cultural evolution or epigenetics (Lewens, 2015; Laland and Brown, in press; Richerson, in press). The causal essentialist view has complementary strengths and weaknesses. It does not put evolution, broadly or narrowly construed, in the foreground, but the causal essentialist account does identify human nature with a manageable number of properties that most people, including philosophers and evolutionary psychologists, would judge to be important—with underlying, causal mechanisms that generate observable behavior.

The evidence surveyed in this book suggests that a hybrid of the nomological and causal essentialist accounts, "evolutionary causal essentialism," can combine their strengths and overcome their weaknesses, allowing the concept of human nature to capture more accurately what scientists have discovered to date, and what they plan to investigate in the future (Heyes, in press, b). This hybrid says that human nature is *the set of mechanisms that underlie the manifestation of species-typical cognitive and behavioral regularities, which humans tend to possess as a result of the evolution of their species*. Crucially, it understands "evolution" to encompass all selection-based evolutionary processes—genetic, epigenetic, and cultural.

Evolutionary causal essentialism inherits some significant assets from its parents. Like the nomological view, it is evolutionary, but it does not commit the essentialist sins found by evolutionists in older theories of human nature (Hull, 1986). For example, it does not offer necessary and sufficient conditions for being human, or imply that human nature is either normative or fixed. Similarly, evolutionary causal essentialism, like causal essentialism, preserves the causal-explanatory function of human nature. Following Aristotle, Locke, and Hume, it casts the components of human nature as underlying, or "hidden," entities that explain obvious differences between humans and other creatures (Samuels, 2012). Evolutionary causal essentialism also escapes some of the problems encountered by its parents. For example, the nomological view has been accused of excessive liberalism because it allows Sikhism and skiing to be components of human nature (Lewens, 2015). In contrast, evolutionary causal essentialism, like plain causal essentialism, would not admit Sikhism and skiing, however widespread they became, because they are sets of manifest characteristics (behaviors, beliefs, values), rather than generative mechanisms of the kind studied by cognitive science.

A more serious charge against the nomological view is that it makes the arbitrary assumption that a characteristic can be a component of human nature if it is a product of the genetic evolution of our species, but not if it results from cultural evolution (Laland and Brown, in press; Lewens, 2015; Richerson, in press). Evolutionary causal essentialism overcomes this problem by specifying that "the evolution of our species" encompasses all selection-based evolutionary processes—genetic, epigenetic, and cultural. The evidence reviewed in this book provides strong motivation for this way of conceptualizing the evolution of our species. Previous research has shown that many, distinctively human *behaviors* have been shaped by cultural evolution, but the evidence reviewed here suggests that the cognitive mechanisms generating those behaviors are also products of selection operating on cultural variants. Cultural evolution—in the strong, selectionist sense (see Chapter 2)—has produced the hidden causes, not just surface manifestations, of human distinctiveness.

The most important implication of the evolutionary causal essentialist view is that human nature is labile; it changes over historical, rather than geological, time. The first signs of literacy date from about six thousand years ago, and now the cognitive gadgets that enable people to read are present in more than 80 to 90 percent of humans. They are underlying, behavior-causing mechanisms that are typical of our species and, therefore, part of human nature. In this sense, and many others, literacy in general, and some specific versions of literacy (for example, Arabic script), have been hugely successful, but success on any scale is likely to be rare. The chances are that a huge number of "start-up" cognitive processes—variants of the types we see in contemporary humans, and wholly different types—have been sustained by cultural evolution for a short time in a restricted area, and then died out long before they became typical of the species. Some cognitive variants will have failed to spread, or become extinct, because they were only locally adaptive. This is almost certainly true of many of the cognitive mechanisms underlying the skills of medieval guild members, such as cordwaining and candle making (see Chapter 1). Other cognitive variants may have had global potential but died out because the groups in which they were evolving were decimated by war or epidemic. Perhaps, for example, cognitive mechanisms that enable humans to learn many languages rapidly in adulthood (Evans, 2013), to remember events with extraordinary precision, or to predict behavior more effectively than any current theory of mind have flourished in particular places at particular times, and then disappeared as a consequence of conflict or misfortune.

The cognitive mechanisms that have "made it"—the ones that are currently part of human nature—are not immune to these dangers. The cognitive instinct view implies that human nature is relatively invulnerable to catastrophe. In a decimated and isolated human population, cultural grist would be lost; the group would lose some of its knowledge and skills. However, with each birth, there would be a new child equipped with Big Special cognitive instincts, ready to innovate and learn from others the knowledge and skills that remained. In contrast, the cognitive gadgets view implies that both grist and mills would be lost. In a skeletal, traumatized population, children would be unlikely to develop the Big Special cognitive mechanisms, such as causal understanding, episodic memory, imitation, and mindreading. Their genetic starter kit (Chapter 3)—social tolerance, social motivation, super-charged associative learning, and executive control—would put them in a better position than chimpanzees to acquire what remains of both grist and mills, but there would be no guarantees. The capacity for cultural evolution, as well as the products of cultural evolution, could be lost.

This is, to say the least, a sobering thought, but the cognitive gadgets view of human nature also has an upside. High Church evolutionary psychology, the cognitive instinct view, famously implies that "our modern skulls house a Stone Age mind" (Cosmides and Tooby, 1997); in other words, the cognitive processes with which we tackle contemporary life were shaped by genetic evolution to meet the needs of small, nomadic bands of people, who devoted most of their energy to digging up plants and hunting animals. In contrast, cultural evolutionary psychology, the cognitive gadgets theory, suggests that distinctively human cognitive mechanisms are light on their feet, constantly changing to meet the demands of new social and physical environments. If that is correct, we need not fear that our minds will be stretched too far by living conditions that depart ever farther from those of hunter-gatherer societies. On the cognitive gadgets view, rather than taxing an outdated mind, new technologies—social media, robotics, virtual reality—merely provide the stimulus for further cultural evolution of the human mind.

CULTURAL EVOLUTIONARY PSYCHOLOGY

Cognitive gadgets theory opens up a "third way." It suggests that distinctively human cognitive mechanisms are adaptive because they are shaped primarily by cultural evolution, not by genetic evolution or intelligent design. This is a bold, testable hypothesis. I have tried in this book to make the hypothesis clear and plausible, but I have no illusions that the case is already conclusive. A great deal more work is needed to test the cognitive gadgets theory and, through the lens of cultural evolutionary psychology, to develop a deeper understanding of the origins and operating characteristics of human minds.

Among the cognitive mechanisms examined in the case studies, selective social learning provides the freshest opportunity for dedicated empirical research. For example, there is plenty of evidence that children show selective social learning from an early age. To test the cognitive gadgets theory, we need experiments that tell us when the selectivity begins to be metacognitive—of a kind that could *not* be produced by domain- and taxon-general attentional mechanisms. We also need to find out whether variation in the content and developmental timetable of metacognitive social learning strategies—within and between cultures—represents poverty or wealth of the stimulus. Does it correlate not merely with children's first-hand experience of "who knows," but with their opportunities to learn from others when to rely on social learning, and who to trust?

Beyond the case studies, there are many other distinctively human cognitive mechanisms to explore. This book has focused on four kinds of cultural learning: selective social learning, imitation, mindreading, and language. A fifth type, normative or moral thinking, is ripe for similar analysis, encompassing behavioral and neurophysiological data from children, adults, and nonhuman animals. What role has cultural evolution played in shaping our capacity to make inferences about what we and others *ought* to do and think—a capacity that not only promotes cooperation, but encourages fidelity of cultural inheritance? Given the evidence, reviewed in this book, that other kinds of cultural learning are cognitive gadgets, one might expect the same to be true of normativity. However, the question is more open for social cognitive mechanisms, such as face processing, that do not appear to be specialized for cultural inheritance. There can be no doubt that input from social sources—the sight of many faces—is necessary for the development of face processing, but it seems unlikely that we learn from others how to recognize faces. Maybe other people play the same role in the development of face recognition as wines play in the development of wine recognition; their faces must be seen, but they don't teach us how to do it, or show us how it should be done. What about asocial, or domain-general, cognitive mechanisms such as causal understanding, episodic memory, and reasoning? I see signs that each of these capacities is a cognitive gadget—signs that it depends on cognitive mechanisms constructed in the course of development through social interaction—but those signs are, for the moment, merely tantalizing. Thorough, wide-ranging analysis of the available data, and dedicated experiments, are needed to test each distinctively human cognitive mechanism for "gadgetry."

Looking further afield, there is the exciting prospect of relating cognition and emotion within the gadgets framework. Barrett's theory of constructed emotion (2017) makes a strong case for the importance of social interaction in shaping human emotions. Could that theory, and others like it (Greenwood, 2015), be enriched by a cultural evolutionary perspective? Are there emotional gadgets? Could we develop a deeper understanding of cognitive gadgets by examining how they have co-evolved, culturally, with distinctively human emotions? Morality is likely to be particularly fertile ground to pursue these questions, given the intimate relationship between normative thinking and distinctively human emotions such as shame and guilt.

One of the advantages of cultural evolutionary psychology is that it brings into sharp focus, and makes tractable, questions about how each cognitive mechanism is put together over time. Distinctively human cognitive mechanisms are complex, adaptive, and adaptable machines, not the kinds of entities that spring into existence fully formed. Therefore, a full explanation of their origins, whether it casts them as cognitive instincts or cognitive gadgets, would include an account of the construction process. It would explain, for example, which "old parts" have been incorporated in the design, how they relate to any new parts, the order in which the parts are assembled, and what kind of environmental input drives the construction process (Chapter 6 did this for imitation). These are important questions, whether construction occurs on a phylogenetic or ontogenetic timescale, over eons or within lifetimes, but they are rarely addressed by nativists. Evolutionary psychologists tend to assume that, if something is a cognitive instinct, it is the responsibility of some other discipline—perhaps genetics or paleo-archaeology, but not cognitive science—to explain how it was constructed (Samuels, 2004). In contrast, cultural evolutionary psychology encourages cognitive scientists and others to develop and test theories about how cognitive gadgets are put together over time. Furthermore, because cultural evolution is faster than genetic evolution, and much of the construction process occurs within lifetimes, the cognitive gadgets theory makes questions about construction empirically tractable. They can be addressed, in collaboration with historians and anthropologists, by research involving contemporary and historical populations, as well as those for which we have only archaeological evidence. We don't have to guess how cognitive mechanisms were put together by genetic evolution in the Pleistocene past. Through laboratory experiments and field studies, we can watch them being constructed in people alive today.

Cultural evolutionary psychology not only raises new questions about distinctive human cognitive mechanisms and their relationships with human emotion, it also suggests that we need to apply new standards of evidence when inquiring about the origins of human minds. If the adaptiveness of cognitive mechanisms can be due not only to genetic evolution and intelligent design, but to cultural selection, the mere fact that a mechanism is adaptive, or difficulty in imagining how it could be learned (intelligent design), does not amount to evidence that the mechanism was shaped by genetic evolution. To be convinced that a mechanism is a cognitive instinct, we need positive evidence of genetic involvement. For example, we need studies showing that identical twins are more alike than fraternal twins, that the mechanism is difficult to acquire in unnatural contexts, or that the acquisition processes do not conform to domain-general principles of learning. "It's in our genes" cannot be used as a default explanation.

CONCLUSION

The framework introduced in this book, cultural evolutionary psychology, combines the strengths of evolutionary psychology and cultural evolutionary theory to answer the question: What makes us such peculiar animals? Like evolutionary psychology at its best, cultural evolutionary psychology takes this to be a question about the mind, drawing on cognitive science, rather than folk psychology, for information about how the mind works. Like cultural evolutionary theory, cultural evolutionary psychology embraces the evidence that human phenotypes are shaped not only by genetic inheritance and learning, but also by cultural evolution. However, unlike both of its conceptual ancestors, cultural evolutionary psychology finds evidence—in social cognitive neuroscience and a broad range of other fields—that the influence of cultural evolution is not confined to the grist of human thought. It has also shaped the mills. Distinctively human cognitive processes are products of cultural group selection. They are not cognitive instincts, but cognitive gadgets.

References

Alem, S., Perry, C. J., Zhu, X., Loukola, O. J., Ingraham, T., Søvik, E., and Chittka, L. (2016). Associative mechanisms allow for social learning and cultural transmission of string pulling in an insect. *PLoS Biology, 14*(10), e1002564.

Alvarez, J. A., and Emory, E. (2006). Executive function and the frontal lobes: A meta-analytic review. *Neuropsychology Review, 16*(1), 17–42.

Amundson, R. (1989). The trials and tribulations of selectionist explanations. In K. Hahlweg and C. A. Hooker (eds.), *Issues in Evolutionary Epistemology.* Albany, NY: State University of New York Press, 413–432.

Anderson, M. L. (2008). Circuit sharing and the implementation of intelligent systems. *Connection Science, 20*(4), 239–251.

Anderson, M. L. (2010). Neural reuse: A fundamental organizational principle of the brain. *Behavioral and Brain Sciences, 33*(4), 245–266.

Anderson, M. L. (2016). After phrenology: Neural reuse and the interactive brain (précis). *Behavioral and Brain Sciences, 39,* e120.

Anderson, M. L., and Finlay, B. L. (2014). Allocating structure to function: The strong links between neuroplasticity and natural selection. *Frontiers in Human Neuroscience, 7,* 918.

Anisfeld, M. (1979). Interpreting "imitative" responses in early infancy. *Science, 205,* 214–215.

Anisfeld, M. (2005). No compelling evidence to dispute Piaget's timetable of the development of representational imitation in infancy. In S. Hurley and

N. Chater (eds.), *Perspectives on Imitation: From Cognitive Neuroscience to Social Science* (Vol. 2). Cambridge, MA: MIT Press, 107–131.

Apperly, I. (2010). *Mindreaders: The Cognitive Basis of "Theory of Mind."*. Hove, UK: Psychology Press.

Apps, M. A., Lesage, E., and Ramnani, N. (2015). Vicarious reinforcement learning signals when instructing others. *Journal of Neuroscience, 35*(7), 2904–2913.

Aquinas, T. (1272). *Summa Theologica* (new ed., 2015). Roccasecca, Italy: Xist Publishing.

Atran, S. (2001). The trouble with memes. *Human Nature, 12*(4), 351–381.

Baer, D. M., and Sherman, J. A. (1964). Reinforcement control of generalized imitation in young children. *Journal of Experimental Child Psychology, 1*(1), 37–49.

Bahrami, B., Olsen, K., Bang, D., Roepstorff, A., Rees, G., and Frith, C. (2012). Together, slowly but surely: The role of social interaction and feedback on the build-up of benefit in collective decision-making. *Journal of Experimental Psychology: Human Perception and Performance, 38*(1), 3–8.

Baillargeon, R., Scott, R. M., and He, Z. (2010). False-belief understanding in infants. *Trends in Cognitive Sciences, 14*(3), 110–118.

Bandura, A., Ross, D., and Ross, S. A. (1963). A comparative test of the status envy, social power, and secondary reinforcement theories of identificatory learning. *Journal of Abnormal and Social Psychology, 67*(6), 527–534.

Bardi, L., Regolin, L., and Simion, F. (2011). Biological motion preference in humans at birth: Role of dynamic and configural properties. *Developmental Science, 14*(2), 353–359.

Bardi, L., Regolin, L., and Simion, F. (2014). The first time ever I saw your feet: Inversion effect in newborns' sensitivity to biological motion. *Developmental Psychology, 50*(4), 986–993.

Barkow, J. H., Cosmides, L., and Tooby, J. (eds.) (1992). *The Adapted Mind: Evolutionary Psychology and the Generation of Culture.* Oxford: Oxford University Press.

Baron-Cohen, S. (1997). *Mindblindness: An Essay on Autism and Theory of Mind.* Cambridge, MA: MIT Press.

Barrett, L. F. (2017). *How Emotions are Made: The Secret Life of the Brain.* Boston, MA: Houghton Mifflin Harcourt.

Barton, R. A., and Venditti, C. (2013). Human frontal lobes are not relatively large. *Proceedings of the National Academy of Sciences, 110*(22), 9001–9006.

Bates, E., and MacWhinney, B. (1982). Functionalist approaches to grammar. In L. Gleitman and E. Wanner (eds.), *Language Acquisition: The State of the Art.* Cambridge: Cambridge University Press.

Bates, E., and MacWhinney, B. (1989). Functionalism and the competition model. In B. MacWhinney and E. Bates (eds.), *The Cross-Linguistic Study of Sentence Processing*. Cambridge: Cambridge University Press.

Battesti, M., Pasquaretta, C., Moreno, C., Teseo, S., Joly, D., Klensch, E., . . . and Mery, F. (2015). Ecology of information: Social transmission dynamics within groups of non-social insects. *Proceedings of the Royal Society of London, Series B: Biological Sciences, 282*(1801), 20142480.

Behrens, T. E., Hunt, L. T., Woolrich, M. W., and Rushworth, M. F. (2008). Associative learning of social value. *Nature, 456*(7219), 245–249.

Behrens, T. E., Woolrich, M. W., Walton, M. E., and Rushworth, M. F. (2007). Learning the value of information in an uncertain world. *Nature Neuroscience, 10,* 1214–1221.

Beran, M. J., Perdue, B. M., Futch, S. E., Smith, J. D., Evans, T. A., and Parrish, A. E. (2015). Go when you know: Chimpanzees' confidence movements reflect their responses in a computerized memory task. *Cognition, 142,* 236–246.

Beran, M. J., Smith, J. D., and Perdue, B. M. (2013). Language-trained chimpanzees (Pan troglodytes) name what they have seen but look first at what they have not seen. *Psychological Science, 24*(5), 660–666.

Berwick, R. C., and Chomsky, N. (2015). *Why Only Us: Language and Evolution.* Cambridge, MA: MIT press.

Bird, G., and Heyes, C. (2005). Effector-dependent learning by observation of a finger movement sequence. *Journal of Experimental Psychology: Human Perception and Performance, 31*(2), 262–275.

Birdsong, D., and Molis, M. (2001). On the evidence for maturational constraints in second-language acquisition. *Journal of Memory and Language, 44*(2), 235–249.

Blackmore, S. (2000). *The Meme Machine.* Oxford: Oxford Paperbacks.

Blakemore, S. J. (2008). The social brain in adolescence. *Nature Reviews Neuroscience, 9*(4), 267–277.

Blakemore, S. J., Winston, J., and Frith, U. (2004). Social cognitive neuroscience: Where are we heading? *Trends in Cognitive Sciences, 8*(5), 216–222.

Blass, E. M., Ganchrow, J. R., and Steiner, J. E. (1984). Classical conditioning in newborn humans 2–48 hours of age. *Infant Behavior and Development, 7*(2), 223–235.

Bloom, P. (2000). *How Children Learn the Meanings of Words.* Cambridge, MA: The MIT Press.

Bloom, P. (2001). How children learn the meanings of words (précis). *Behavioral and Brain Sciences, 24*(6), 1095–1103.

Boakes, R. (1984). *From Darwin to Behaviourism: Psychology and the Minds of Animals.* CUP Archive.

Bock, J., Poeggel, G., Gruss, M., Wingenfeld, K., and Braun, K. (2014). Infant cognitive training preshapes learning-relevant prefrontal circuits for adult learning: Learning-induced tagging of dendritic spines. *Cerebral Cortex, 24*(11), 2920–2930.

Boeckx, C. (2006). *Linguistic Minimalism: Origins, Concepts, Methods, and Aims.* Oxford: Oxford University Press.

Bohannon, J. N., MacWhinney, B., and Snow, C. (1990). No negative evidence revisited: Beyond learnability or who has to prove what to whom? *Developmental Psychology, 26*(2), 221–226.

Bohannon, J. N., and Stanowicz, L. B. (1988). The issue of negative evidence: Adult responses to children's language errors. *Developmental Psychology, 24*(5), 684–689.

Boogert, N. J., Giraldeau, L-A., and Lefebvre, L. (2008). Song complexity correlates with learning ability in zebra finch males. *Animal Behaviour, 76,* 1735–1741.

Bornkessel-Schlesewsky, I., Schlesewsky, M., Small, S. L., and Rauschecker, J. P. (2015). Neurobiological roots of language in primate audition: Common computational properties. *Trends in Cognitive Sciences, 19*(3), 142–150.

Bouchard, J., Goodyer, W., and Lefebvre, L. (2007). Social learning and innovation are positively correlated in pigeons. *Animal Cognition, 10,* 259–266.

Bouchard, T. J. (2014). Genes, evolution and intelligence. *Behavior Genetics, 44*(6), 549–577.

Boyd, R., and Richerson, P. J. (1988). *Culture and the Evolutionary Process.* Chicago: University of Chicago Press.

Brem, S., Bucher, K., Halder, P., Summers, P., Dietrich, T., Martin, E., and Brandeis, D. (2006). Evidence for developmental changes in the visual word processing network beyond adolescence. *NeuroImage, 29*(3), 822–837.

Brown, R., and Hanlon, C. (1970). Derivational complexity and order of acquisition in child speech. *Cognition and the Development of Language.* New York: Wiley.

Brown, R. L. (2014). Identifying behavioral novelty. *Biological Theory, 9*(2), 135–148.

Brusse, C. (2017). Making do without selection—review essay of "Cultural Evolution: Conceptual Challenges" by Tim Lewens. *Biology and Philosophy, 32*(2), 307–319.

Bull, R., Phillips, L. H., and Conway, C. A. (2008). The role of control functions in mentalizing: Dual-task studies of theory of mind and executive function. *Cognition, 107*(2), 663–672.

Burkart, J. M., Hrdy, S. B., and Van Schaik, C. P. (2009). Cooperative breeding and human cognitive evolution. *Evolutionary Anthropology: Issues, News, and Reviews, 18*(5), 175–186.

Burnham, D., Kitamura, C., and Vollmer-Conna, U. (2002). What's new, pussycat? On talking to babies and animals. *Science, 296*(5572), 1435–1435.

Burrow, C. (1993). *Epic Romance: Homer to Milton.* Oxford: Oxford University Press.

Butterfill, S. A., and Apperly, I. A. (2013). How to construct a minimal theory of mind. *Mind and Language, 28*(5), 606–637.

Butterfill, S., Apperly, I., Rakoczy, H., Spaulding, S., and Zawidzki, T. (2013). Symposium on S. Butterfill and I. Apperly, "How to Construct a Minimal Theory of Mind." Mind and Language Symposia at the Brains Blog. http://philosophyofbrains.com/2013/11/11/symposium-on-butterfill-and-apperlys-how-to-construct-a-minimal-theory-of-mind-mind-language-28-5-606-63.aspx

Byrne, R. W., and Rapaport, L. G. (2011). What are we learning from teaching? *Animal Behaviour, 82*(5), 1207–1211.

Caldwell, C. A., Atkinson, M., and Renner, E. (2016). Experimental approaches to studying cumulative cultural evolution. *Current Directions in Psychological Science, 25*(3), 191–195.

Calvo-Merino, B., Glaser, D. E., Grèzes, J., Passingham, R. E., and Haggard, P. (2005). Action observation and acquired motor skills: An fMRI study with expert dancers. *Cerebral Cortex, 15*(8), 1243–1249.

Calvo-Merino, B., Grèzes, J., Glaser, D. E., Passingham, R. E., and Haggard, P. (2006). Seeing or doing? Influence of visual and motor familiarity in action observation. *Current Biology, 16*(19), 1905–1910.

Campbell, D. T. (1965). Variation and selective retention in socio-cultural evolution. *Social Change in Developing Areas, 19,* 26–27.

Campbell, D. T. (1974). Evolutionary epistemology. In P.A. Schlipp (ed.), *The Philosophy of Karl Popper.* LaSalle, IL: Open Court, 413–463.

Cardoso, C., Ellenbogen, M. A., and Linnen, A. M. (2014). The effect of intranasal oxytocin on perceiving and understanding emotion on the Mayer–Salovey–Caruso emotional intelligence test (MSCEIT). *Emotion, 1,* 43–50.

Carey, S. (2009). *The Origin of Concepts.* Oxford: Oxford University Press.

Caro, T. M., and Hauser, M. D. (1992). Is there teaching in nonhuman animals? *The Quarterly Review of Biology, 67*(2), 151–174.

Carpendale, J. I., and Lewis, C. (2004). Constructing an understanding of mind: The development of children's social understanding within social interaction. *Behavioral and Brain Sciences, 27*(1), 79–96.

Carpenter, M., and Call, J. (2013). How joint is the joint attention of apes and human infants? In J. Metcalfe and H.S. Terrace (eds.), *Agency and Joint Attention.* New York: Oxford University Press, 49–61.

Caselli, L., and Chelazzi, L. (2011). Does the macaque monkey provide a good model for studying human executive control? A comparative behavioral study of task switching. *PLoS One, 6*(6), e21489.

Caspers, S., Zilles, K., Laird, A. R., and Eickhoff, S. B. (2010). ALE meta-analysis of action observation and imitation in the human brain. *NeuroImage, 50*(3),1148–1167.

Catmur, C., Mars, R. B., Rushworth, M. F., and Heyes, C. (2011). Making mirrors: Premotor cortex stimulation enhances mirror and counter-mirror motor facilitation. *Journal of Cognitive Neuroscience, 23*(9), 2352–2362.

Catmur, C., Press, C., and Heyes, C. (2016). Mirror associations. In R. A. Murphy and R. C. Honey (eds.), *The Wiley Handbook of the Cognitive Neuroscience of Learning.* Hoboken, NJ: Wiley.

Catmur, C., Santiesteban, I., Conway, J. R., Heyes, C., and Bird, G. (2016). Avatars and arrows in the brain. *NeuroImage, 132,* 8–10.

Catmur, C., Walsh, V., and Heyes, C. (2007). Sensorimotor learning configures the human mirror system. *Current Biology, 17*(17), 1527–1531.

Catmur, C., Walsh, V., and Heyes, C. (2009). Associative sequence learning: The role of experience in the development of imitation and the mirror system. *Philosophical Transactions of the Royal Society of London, Series B: Biological Sciences, 364*(1528), 2369–2380.

Cavalli-Sforza, L. L. (1997). Genes, peoples, and languages. *Proceedings of the National Academy of Sciences, 94,* 7719–7724.

Cavalli-Sforza, L. L., and Feldman, M. W. (1981). *Cultural Transmission and Evolution: A Quantitative Approach.* Princeton, NJ: Princeton University Press.

Chang, E. F., and Merzenich, M. M. (2003). Environmental noise retards auditory cortical development. *Science, 300*(5618), 498–502.

Changizi, M. A., Zhang, Q., Ye, H., and Shimojo, S. (2006). The structures of letters and symbols throughout human history are selected to match those found in objects in natural scenes. *The American Naturalist, 167*(5), E117–E139.

Chater, N., and Christiansen, M. H. (2010). Language acquisition meets language evolution. *Cognitive Science, 34*(7), 1131–1157.

Chater, N. and Heyes, C. (1994). Animal concepts: Content and discontent. *Mind and Language, 9*(3), 209–246.

Chater, N., Reali, F., and Christiansen, M. H. (2009). Restrictions on biological adaptation in language evolution. *Proceedings of the National Academy of Sciences, 106*(4), 1015–1020.

Chomsky, N. (1957). *Syntactic Structures.* The Hague: Mouton and Co.

Chomsky, N. (1965). *Aspects of the Theory of Syntax.* Cambridge, MA: MIT Press.

Chomsky, N. (1972). *Language and Mind.* New York: Foris.

Chomsky, N. (1981). *Lectures on Government and Binding.* New York: Foris.

Chomsky, N. (1988). *Language and Problems of Knowledge: The Managua Lectures* (Vol. 16). Cambridge, MA: MIT Press.

Chomsky, N. (1995). Categories and transformations. In *The Minimalist Program, 219–394.*

Christiansen, M. H., and Chater, N. (2008). Language as shaped by the brain. *Behavioral and Brain Sciences, 31*(5), 489–509.

Christiansen, M. H., and Chater, N. (2016). *Creating Language: Integrating Evolution, Acquisition, and Processing.* Cambridge, MA: MIT Press.

Christiansen, M. H., and MacDonald, M. C. (2009). A usage-based approach to recursion in sentence processing. *Language Learning, 59*(s1), 126–161.

Churchland, P. S., and Winkielman, P. (2012). Modulating social behavior with oxytocin: How does it work? What does it mean? *Hormones & Behavior, 61,* 392–399.

Cieri, R. L., Churchill, S. E., Franciscus, R. G., Tan, J., Hare, B., Athreya, S., . . . and Wrangham, R. (2014). Craniofacial feminization, social tolerance, and the origins of behavioral modernity. *Current Anthropology, 55*(4), 419–443.

Clarke, E., and Heyes, C. (2017). The swashbuckling anthropologist: Henrich on the secret of our success. *Biology and Philosophy, 32,* 289–305.

Collins, A., and Koechlin, E. (2012). Reasoning, learning, and creativity: Frontal lobe function and human decision-making. *PLoS Biology, 10*(3), e1001293.

Coltheart, M., Rastle, K., Perry, C., Langdon, R., and Ziegler, J. (2001). DRC: A dual route cascaded model of visual word recognition and reading aloud. *Psychological Review, 108*(1), 204–256.

Cook, J. L. (2014). Task-relevance dependent gradients in medial prefrontal and temporoparietal cortices suggest solutions to paradoxes concerning self / other control. *Neuroscience and Biobehavioral Reviews, 42C,* 298–302.

Cook, M., Mineka, S., Wolkenstein, B., and Laitsch, K. (1985). Observational conditioning of snake fear in unrelated rhesus monkeys. *Journal of Abnormal Psychology, 93,* 355–372.

Cook, R., Bird, G., Catmur, C., Press, C., and Heyes, C. (2014). Mirror neurons: From origin to function. *Behavioral and Brain Sciences, 37*(2), 177–192.

Cook, R., Dickinson, A., and Heyes, C. (2012). Contextual modulation of mirror and countermirror sensorimotor associations. *Journal of Experimental Psychology: General, 141*(4), 774–787.

Cook, R., Johnston, A., and Heyes, C. (2013). Facial self-imitation: Objective measurement reveals no improvement without visual feedback. *Psychological Science, 24*(1), 93–98.

Cook, R., Press, C., Dickinson, A., and Heyes, C. (2010). Is the acquisition of automatic imitation sensitive to sensorimotor contingency? *Journal of Experimental Psychology: Human Perception and Performance, 36,* 840–852.

Cook, R. G., Brown, M. F., and Riley, D. A. (1985). Flexible memory processing by rats: Use of prospective and retrospective information in the radial maze. *Journal of Experimental Psychology: Animal Behavior Processes, 11*(3), 453–469.

Cooper, R. P., and Aslin, R. N. (1990). Preference for infant-directed speech in the first month after birth. *Child Development, 61*(5), 1584–1595.

Cooper, R. P., Cook, R., Dickinson, A., and Heyes, C. (2013). Associative (not Hebbian) learning and the mirror neuron system. *Neuroscience Letters, 540,* 28–36.

Corbeill, A. (2004). *Nature embodied: Gesture in Ancient Rome.* Princeton, NJ: Princeton University Press.

Cosmides, L., and Tooby, J. (1994). Beyond intuition and instinct blindness: Toward an evolutionary rigorous cognitive science. *Cognition, 50,* 41–77.

Cosmides, L., and Tooby, J. (1997). *Evolutionary Psychology: A Primer.* https://www.psych.ucsb.edu/research/cep/primer.html.

Coussi-Korbel, S., and Fragaszy., D. M. (1995). On the relation between social dynamics and social learning. *Animal Behaviour, 50,* 1441–1453.

Cowie, F. (2016). Innateness and language. In E. N. Zalta (ed.), *The Stanford Encyclopedia of Philosophy.* https://plato.stanford.edu/archives/win2016/entries/innateness-language.

Crain, S., Goro, T., and Thornton, R. (2006). Language acquisition is language change. *Journal of Psycholinguistic Research, 35*(1), 31–49.

Cronk, L. (1991). Human behavioral ecology. *Annual Review of Anthropology, 20*(1), 25–53.

Cross, E. S., Hamilton, A. F., and Grafton, S. T. (2006). Building a motor simulation de novo: Observation of dance by dancers. *NeuroImage, 31*(3),1257–1267.

Csibra, G., and Gergely, G. (2006). Social learning and social cognition: The case for pedagogy. In Y. Munakata and M. H. Johnson (eds.), *Processes of Change in Brain and Cognitive Development* (Vol. 21, Attention and Performance). Oxford: Oxford University Press, 249–274.

Culicover, P. W. (1999). *Syntactic Nuts: Hard Cases, Syntactic Theory, and Language Acquisition* (Vol. 1). Oxford University Press on Demand.

Culicover, P. W., and Jackendoff, R. (2005). *Simpler Syntax.* Oxford: Oxford University Press.

Dąbrowska, E. (2012). Different speakers, different grammars: Individual differences in native language attainment. *Linguistic Approaches to Bilingualism, 2*(3), 219–253.

Damuth, J., and Heisler, I. L. (1988). Alternative formulations of multilevel selection. *Biology and Philosophy, 3*(4), 407–430.

Darwin, C. (1868). *The Variation in Animals and Plants under Domestication.* London: John Murray.

Darwin, C. (1871). *The Descent of Man, and Selection in Relation to Sex.* London: John Murray.

Dawkins, R. (1989). *The Selfish Gene.* Oxford: Oxford University Press.

Dawson, E. H., Avarguès-Weber, A., Chittka, L., and Leadbeater, E. (2013). Learning by observation emerges from simple associations in an insect model. *Current Biology, 23*(8), 727–730.

De Casper, A. J., and Fifer, W. P. (1980). Newborn preference for the maternal voice: An indication of early attachment. Paper delivered at a meeting of the Southeastern Conference on Human Development, Alexandria, VA.

De Casper, A. J., and Spence, M. J. (1986). Prenatal maternal speech influences newborns' perception of speech sounds. *Infant Behavior and Development, 9*(2), 133–150.

Dehaene, S. (2009). *Reading in the Brain: The New Science of How We Read.* London: Penguin.

Dehaene, S. (2014). Reading in the brain revised and extended: Response to comments. *Mind and Language, 29*(3), 320–335.

Dehaene, S., and Cohen, L. (2011). The unique role of the visual word form area in reading. *Trends in Cognitive Sciences, 15*(6), 254–262.

Dehaene, S., Pegado, F., Braga, L. W., Ventura, P., Nunes Filho, G., Jobert, A., Dehaene-Lambertz, G., Kolinsky, R., Morais, J., and Cohen, L. (2010). How learning to read changes the cortical networks for vision and language. *Science, 330*(6009), 1359–1364.

De Klerk, C. C., Johnson, M. H., Heyes, C., and Southgate, V. (2015). Baby steps: Investigating the development of perceptual-motor couplings in infancy. *Developmental Science, 18*(2), 270–280.

Della Libera, C., and Chelazzi, L. (2006). Visual selective attention and the effects of monetary rewards. *Psychological Science, 17*(3), 222–227.

Demetras, M. J., Post, K. N., and Snow, C. E. (1986). Feedback to first language learners: The role of repetitions and clarification questions. *Journal of Child Language, 13*(2), 275–292.

Dennett, D. C. (1969). *Content and Consciousness*. London: Routledge.

Dennett, D. C. (1984). Cognitive wheels: The frame problem of AI. In C. Hookway (ed.), *Minds, Machines and Evolution*. Cambridge: Cambridge University Press, 129–151.

Dennett, D. C. (1987). *The Intentional Stance*. Cambridge, MA: MIT Press.

Dennett, D. C. (1990). Memes and the exploitation of imagination. *The Journal of Aesthetics and Art Criticism, 48*(2), 127–135.

Dennett, D. C. (1991). *Consciousness Explained*. Boston: Little, Brown and Co.

Dennett, D. C. (2001). The evolution of culture. *The Monist, 84*(3), 305–324.

Den Ouden, H. E., Friston, K. J., Daw, N. D., McIntosh, A. R., and Stephan, K. E. (2009). A dual role for prediction error in associative learning. *Cerebral Cortex, 19*(5), 1175–1185.

De Villiers, P. A., and de Villiers, J. G. (2012). Deception dissociates from false belief reasoning in deaf children: Implications for the implicit versus explicit theory of mind distinction. *British Journal of Developmental Psychology, 30*(1), 188–209.

Diaconescu, A. O., Mathys, C., Weber, L. A., Daunizeau, J., Kasper, L., Lomakina, E. I., . . . and Stephan, K. E. (2014). Inferring on the intentions of others by hierarchical Bayesian learning. *PLoS Computational Biology, 10*(9), e1003810.

Diamond, A. (2013). Executive functions. *Annual Review of Psychology, 64*, 135–168.

Dickinson, A. (1980). *Contemporary Animal Learning Theory*. Cambridge: Cambridge University Press.

Dickinson, A. (2012). Associative learning and animal cognition. *Philosophical Transactions of the Royal Society of London, Series B: Biological Sciences, 367*(1603), 2733–2742.

Di Giorgio, E., Leo, I., Pascalis, O., and Simion, F. (2012). Is the face-perception system human-specific at birth?. *Developmental Psychology, 48*(4), 1083–1090.

Di Pellegrino, G., Fadiga, L., Fogassi, L., Gallese, V., and Rizzolatti, G. (1992). Understanding motor events: A neurophysiological study. *Experimental Brain Research, 91*(1), 176–180.

Dolk, T., Hommel, B., Colzato, L. S., Schütz-Bosbach, S., Prinz, W., and Liepelt, R. (2011). How "social" is the social Simon effect? *Frontiers in Psychology.* https://doi.org/10.3389/fpsyg.2011.00084.

Dorrance, B. R., and Zentall, T. R. (2001). Imitative learning in Japanese quail depends on the motivational state of the observer quail at the time of observation. *Journal of Comparative Psychology, 115,* 62–67.

Dumontheil, I., Apperly, I. A., and Blakemore, S. J. (2010). Online usage of theory of mind continues to develop in late adolescence. *Developmental Science, 13*(2), 331–338.

Duncan, J., Schramm, M., Thompson, R., and Dumontheil, I. (2012). Task rules, working memory, and fluid intelligence. *Psychonomic Bulletin & Review, 19*(5), 864–870.

Dupierrix, E., de Boisferon, A. H., Méary, D., Lee, K., Quinn, P. C., Di Giorgio, E., . . . and Pascalis, O. (2014). Preference for human eyes in human infants. *Journal of Experimental Child Psychology, 123,* 138–146.

Duranti, A. (2008). Further reflections on reading other minds. *Anthropological Quarterly, 81,* 483–494.

Eco, U. (1995). *The Search for the Perfect Language.* Oxford: Wiley-Blackwell.

Efferson, C., Richerson, P. J., McElreath, R., Lubell, M., Edsten, E., Waring, T. M., Paciotti, B., and Baum, W. (2007). Learning, productivity, and noise: An experimental study of cultural transmission on the Bolivian Alti-plano. *Evolution and Human Behavior, 28,* 11–17.

Emberson, L. L., Richards, J. E., and Aslin, R. N. (2015). Top-down modulation in the infant brain: Learning-induced expectations rapidly affect the sensory cortex at 6 months. *Proceedings of the National Academy of Sciences, 112*(31), 9585–9590.

Eriksson, K. (2012). The nonsense math effect. *Judgment and Decision Making, 7,* 746–749.

Evans, J. S. B., and Stanovich, K. E. (2013). Dual-process theories of higher cognition: Advancing the debate. *Perspectives on Psychological Science, 8*(3), 223–241.

Evans, N. (2013). Multilingualism as the primal human condition: What we have to learn from small-scale speech communities. Keynote address delivered at *9th International Symposium on Bilingualism,* Singapore.

Evans, N., and Levinson, S. C. (2009). The myth of language universals: Language diversity and its importance for cognitive science. *Behavioral and Brain Sciences, 32*(5), 429–448.

Everett, D. L. (2005). Cultural constraints on grammar and cognition in Pirahã. Another look at the design features of human language. *Current Anthropology, 46,* 621–646.

Eysenck, M. W., and Keane, M. T. (2015). *Cognitive Psychology: A Student's Handbook* (7th ed.). Hove, East Sussex: Psychology Press.

Fagot, J., and Cook, R. G. (2006). Evidence for large long-term memory capacities in baboons and pigeons and its implications for learning and the evolution of cognition. *Proceedings of the National Academy of Sciences,* 103(46), 17564–17567.

Falck-Ytter, T., Gredebäck, G., and von Hofsten, C. (2006). Infants predict other people's action goals. *Nature Neuroscience, 9*(7), 878–879.

Farroni, T., Mansfield, E. M., Lai, C., and Johnson, M. H. (2003). Infants perceiving and acting on the eyes: Tests of an evolutionary hypothesis. *Journal of Experimental Child Psychology, 85*(3), 199–212.

Fedorenko, E., Duncan, J., and Kanwisher, N. (2012). Language-selective and domain-general regions lie side by side within Broca's area. *Current Biology, 22*(21), 2059–2062.

Fernald, A. (1991). Prosody in speech to children: Prelinguistic and linguistic functions. *Annals of Child Development, 8,* 43–80.

Fernandes, H. B., Woodley, M. A., and te Nijenhuis, J. (2014). Differences in cognitive abilities among primates are concentrated on g: Phenotypic and phylogenetic comparisons with two meta-analytical databases. *Intelligence, 46,* 311–322.

Ferrari, P. F., Rozzi, S., and Fogassi, L. (2005). Mirror neurons responding to observation of actions made with tools in monkey ventral premotor cortex. *Journal of Cognitive Neuroscience, 17*(2), 212–226.

Fiorito, G., and Scotto, P. (1992). Observational learning in *Octopus vulgaris. Science, 256,* 545–547.

Fitch, W. T., Hauser, M. D., and Chomsky, N. (2005). The evolution of the language faculty: Clarifications and implications. *Cognition, 97,* 179–210.

Flege, J. E., Yeni-Komshian, G. H., and Liu, S. (1999). Age constraints on second-language acquisition. *Journal of Memory and Language, 41*(1), 78–104.

Fleming, S. M., Dolan, R. J., and Frith, C. D. (2012). Metacognition: Computation, biology and function. *Philosophical Transactions of the Royal Society, Series B: Biological Sciences,* 367, 1280–1286.

Floccia, C., Christophe, A., and Bertoncini, J. (1997). High-amplitude sucking and newborns: The quest for underlying mechanisms. *Journal of Experimental Child Psychology, 64*(2), 175–198.

Fogarty, L., Rendell, L., and Laland, K. N. (2012). Mental time travel, memory and the social learning strategies tournament. *Learning & Motivation, 43,* 241–246.

Forssberg, H,. Eliasson, A. C., Kinoshita, H., Johansson, R. S., and Westling, G. (1991). Development of human precision grip. I. Basic coordination of force. *Experimental Brain Research, 85,* 451–457.

Frank, M. C., Vul, E., and Johnson, S. P. (2009). Development of infants' attention to faces during the first year. *Cognition, 110*(2), 160–170.

Frith, C. D., and Frith, U. (2012). Mechanisms of social cognition. *Annual Review of Psychology, 63,* 287–313.

Frith, U. (2001). Mind blindness and the brain in autism. *Neuron, 32*(6), 969–979.

Fusaro, M., and Harris, P. L. (2008). Children assess informant reliability using bystanders' non-verbal cues. *Developmental Science, 11* (5), 771–777.

Gabi, M., Neves, K., Masseron, C., Ribeiro, P. F. M., Ventura-Antunes, L., Torres, L., Mota, B., Kaas, J. H., and Herculano-Houzel, S. (2016). No relative expansion of the number of prefrontal neurons in primate and human evolution. *Proceedings of the National Academy of Sciences,* 201610178.

Galef, B. G. (1971). Social effects in the weaning of domestic rat pups. *Journal of Comparative and Physiological Psychology, 75*(3), 358–362.

Galef, B. G., Dudley, K. E., and Whiskin, E. E. (2008). Social learning of food preferences in "dissatisfied" and "uncertain" Norway rats. *Animal Behaviour, 75,* 631–637.

Gallese, V., Fadiga, L., Fogassi, L., and Rizzolatti, G. (1996). Action recognition in the premotor cortex. *Brain, 119*(2), 593–609.

Gangestad, S. W. (2016). An evolutionary perspective on oxytocin and its behavioral effects. *Current Opinion in Psychology, 7,* 115–119.

Garcia, E., Baer, D. M., and Firestone, I. (1971). The development of generalized imitation within topographically determined boundaries. *Journal of Applied Behavior Analysis, 4*(2), 101–112.

Garvert, M. M., Moutoussis, M., Kurth-Nelson, Z., Behrens, T. E., and Dolan, R. J. (2015). Learning-induced plasticity in medial prefrontal cortex predicts preference malleability. *Neuron, 85*(2), 418–428.

Giles, A., and Rovee-Collier, C. (2011). Infant long-term memory for associations formed during mere exposure. *Infant Behavior and Development, 34*(2), 327–338.

Godfrey-Smith, P. (2009). *Darwinian Populations and Natural Selection.* Oxford: Oxford University Press.

Godfrey-Smith, P. (2012). Darwinism and cultural change. *Philosophical Transactions of the Royal Society, Series B: Biological Sciences, 367*(1599), 2160–2170.

Goldman, A. I. (2006). *Simulating Minds: The Philosophy, Psychology, and Neuroscience of Mindreading.* Oxford: Oxford University Press.

Goldstein, M. H., and Schwade, J. A. (2008). Social feedback to infants' babbling facilitates rapid phonological learning. *Psychological Science, 19*(5), 515–523.

Goodnow, J. J. (1955). Determinants of choice-distribution in two-choice situations. *The American Journal of Psychology, 68*(1), 106–116.

Gopnik, A., and Wellman, H. M. (2012). Reconstructing constructivism: Causal models, Bayesian learning mechanisms, and the theory theory. *Psychological Bulletin, 138*, 1085–1108.

Grafen, A. (1984). Natural selection, kin selection and group selection. In J. R. Krebs and N. B. Davies (eds.), *Behavioural Ecology: An Evolutionary Approach* (2nd ed.). Oxford: Blackwell Scientific, 62–84.

Gray, R. D., and Atkinson, Q. D. (2003). Language-tree divergence times support the Anatolian theory of Indo-European origin. *Nature, 426*(6965), 435–439.

Greenwood, J. (2015). *Becoming Human: The Ontogenesis, Metaphysics, and Expression of Human Emotionality.* Cambridge, MA: MIT Press.

Gregory, R., Cheng, H., Rupp, H. A., Sengelarb, D. R., and Heiman, J. R. (2015). Oxytocin increases VTA activation to infant and sexual stimuli in nulliparous and postpartum women. *Hormones and Behavior, 69*, 82–88.

Grice, H. P. (1957). Meaning. *The Philosophical Review, 66*(3), 377–388.

Griffiths, P. E. (2017). The distinction between innate and acquired characteristics. In E. N. Zalta (ed.), *The Stanford Encyclopedia of Philosophy* (Spring 2017 Edition). https://plato.stanford.edu/archives/spr2017/entries/innate-acquired/.

Griffiths, P. E., Pocheville, A., Calcott, B., Stotz, K., Kim, H., and Knight, R. (2015). Measuring causal specificity. *Philosophy of Science, 82*(4), 529–555.

Gruber, C. W. (2014). Physiology of invertebrate oxytocin and vasopressin neuropeptides. *Experimental Physiology, 99*, 51–55.

Grusec, J. E., and Abramovitch, R. (1982). Imitation of peers and adults in a natural setting: A functional analysis. *Child Development, 53*, 636–642.

Güss, C. D., and Wiley, B. (2007). Metacognition of problem-solving strategies in Brazil, India, and the United States. *Journal of Cognition and Culture, 7,* 1–25.

Guzzon, D., Brignani, D., Miniussi, C., and Marzi, C. A. (2010). Orienting of attention with eye and arrow cues and the effect of overtraining. *Acta Psychologica, 134*(3), 353–362.

Hakuta, K., Bialystok, E., and Wiley, E. (2003). Critical evidence: A test of the critical-period hypothesis for second-language acquisition. *Psychological Science, 14*(1), 31–38.

Harris, W. V. (ed.) (2013). *Mental Disorders in the Classical World.* Leiden: Brill.

Harshaw, C., and Lickliter, R. (2007). Interactive and vicarious acquisition of auditory preferences in Northern bobwhite (*Colinus virginianus*) chicks. *Journal of Comparative Psychology, 121*(3), 320–331.

Harvey, O. J., and Rutherford, J. (1960). Status in the informal group: Influence and influencibility at differing age levels. *Child Development, 31,* 377–385.

Haslinger, B., Erhard, P., Altenmuller, E., Schroeder, U., Boecker, H., and Ceballos-Baumann, A. O. (2005). Transmodal sensorimotor networks during action observation in professional pianists. *Journal of Cognitive Neuroscience, 17*(2), 282–293.

Haun, D. B. M., Rapold, C. J., Call, J., Janzen, G., and Levinson, S. C. (2006). Cognitive cladistics and cultural override in Hominid spatial cognition. *Proceedings of the National Academy of Sciences, 103*(46), 17568–17573.

Hauser, M. D., Chomsky, N., and Fitch, W. T. (2002). The faculty of language: What is it, who has it, and how did it evolve? *Science, 298*(5598), 1569–1579.

Heine, S. J., Kitayama, S., Lehman, D.R., Takata, T., Ide, E., Leung, C., and Matsumoto, H. (2001). Divergent consequences of success and failure in Japan and North America: An investigation of self-improving motivations and malleable selves. *Journal of Personality and Social Psychology, 81,* 599–615.

Henrich, J. (2004). Demography and cultural evolution: How adaptive cultural processes can produce maladaptive losses: The Tasmanian case. *American Antiquity, 69,* 197–214.

Henrich, J. (2015). *The Secret of Our Success: How Culture is Driving Human Evolution, Domesticating Our Species, and Making Us Smarter.* Princeton: Princeton University Press.

Henrich, J., and Broesch, J. (2011). On the nature of cultural transmission networks: Evidence from Fijian villages for adaptive learning biases. *Philosophical Transactions of the Royal Society, Series B: Biological Sciences, 366,* 1139–1148.

Herrnstein, R. J., Loveland, D. H., and Cable, C. (1976). Natural concepts in pigeons. *Journal of Experimental Psychology, 2*(4), 285–302.

Heyes, C. (1993). Imitation, culture and cognition. *Animal Behaviour, 46*(5), 999–1010.

Heyes, C. (1994). Social learning in animals: Categories and mechanisms. *Biological Reviews, 69*(2), 207–231.

Heyes, C. (2003). Four routes of cognitive evolution. *Psychological Review, 110*(4), 713–727.

Heyes, C. (2011). Automatic imitation. *Psychological Bulletin, 137*(3), 463–483.

Heyes, C. (2012a). Grist and mills: On the cultural origins of cultural learning. *Philosophical Transactions of the Royal Society, Series B: Biological Sciences, 367*(1599), 2181–2191.

Heyes, C. (2012b). Simple minds: A qualified defence of associative learning. *Philosophical Transactions of the Royal Society of London, Series B: Biological Sciences, 367*(1603), 2695–2703.

Heyes, C. (2012c). What's social about social learning? *Journal of Comparative Psychology, 126*(2), 193–202.

Heyes, C. (2013). What can imitation do for cooperation? In K. Sterelny, R. Joyce, B. Calcott, and B. Fraser (eds.), *Cooperation and its Evolution.* Cambridge, MA: MIT Press, 313–331.

Heyes, C. (2014a). Submentalizing: I am not really reading your mind. *Perspectives on Psychological Science, 9*(2), 131–143.

Heyes, C. (2014b). False belief in infancy: A fresh look. *Developmental Science, 17*(5), 647–659.

Heyes, C. (2015). Animal mindreading: What's the problem? *Psychonomic Bulletin & Review, 22*(2), 313–327.

Heyes, C. (2016a). Blackboxing: Social learning strategies and cultural evolution. *Philosophical Transactions of the Royal Society, Series B: Biological Sciences, 371*(1693), 20150369.

Heyes, C. (2016b). Born pupils? Natural pedagogy and cultural pedagogy. *Perspectives on Psychological Science, 11*(2), 280–295.

Heyes, C. (2016c). Who knows? Metacognitive social learning strategies. *Trends in Cognitive Sciences, 20*(3), 204–213.

Heyes, C. (2016d). Imitation: Not in our genes. *Current Biology, 26*(10), R412–R414.

Heyes, C. (2016e). Homo imitans? Seven reasons why imitation couldn't possibly be associative. *Philosophical Transactions of the Royal Society, Series B: Biological Sciences, 371*(1686), 20150069.

Heyes, C. (2017a). When does social learning become cultural learning? *Developmental Science, 20*(2), e12350.

Heyes, C. (2017b). Apes submentalise. *Trends in Cognitive Sciences, 21*(1), 1–2.
Heyes, C. (in press, a). Enquire within: Cultural evolution and cognitive science. *Philosophical Transactions of the Royal Society, Series B: Biological Sciences.*
Heyes, C. (in press, b). Human nature, natural pedagogy, and evolutionary causal essentialism. In T. Lewens and E. Hannon (eds.), *Why We Disagree about Human Nature.* Oxford: Oxford University Press.
Heyes, C., Bird, G., Johnson, H., and Haggard, P. (2005). Experience modulates automatic imitation. *Cognitive Brain Research, 22*(2), 233–240.
Heyes, C., and Frith, C. D. (2014). The cultural evolution of mind reading. *Science, 344*(6190), 1243091.
Heyes, C., and Galef, Jr., B. G. (eds.) (1996). *Social Learning in Animals: The Roots of Culture.* San Diego, CA: Academic Press.
Heyes, C., and Pearce, J. M. (2015). Not-so-social learning strategies. *Proceedings of the Royal Society of London, Series B: Biological Sciences, 282*(1802), 20141709.
Heyes, C., and Ray, E. D. (2000). What is the significance of imitation in animals? *Advances in the Study of Behavior, 29,* 215–245.
Hill, M. R., Boorman, E. D., and Fried, I. (2016). Observational learning computations in neurons of the human anterior cingulate cortex. *Nature Communications, 7,* doi:10.1038/ncomms12722.
Holden, C., and Mace, R. (2009). Phylogenetic analysis of the evolution of lactose digestion in adults. *Human Biology, 81*(5 / 6), 597–619.
Holland, P. C. (1992). Occasion setting in Pavlovian conditioning. *Psychology of Learning and Motivation, 28,* 69–125.
Hood, B. M., Willen, J. D., and Driver, J. (1998). Adult's eyes trigger shifts of visual attention in human infants. *Psychological Science, 9*(2), 131–134.
Hoppitt, W., and Laland, K. N. (2013). *Social Learning: An Introduction to Mechanisms, Methods, and Models.* Princeton, NJ: Princeton University Press.
Hove, M. J., and Risen, J. L. (2009). It's all in the timing: Interpersonal synchrony increases affiliation. *Social Cognition, 27*(6), 949–960.
Hrdy, S. B. (2011). *Mothers and Others.* Cambridge, MA: Harvard University Press.
Hsu, H. J., and Bishop, D. V. (2014). Sequence-specific procedural learning deficits in children with specific language impairment. *Developmental Science, 17*(3), 352–365.
Hsu, H. J., Tomblin, J. B., and Christiansen, M. H. (2014). Impaired statistical learning of non-adjacent dependencies in adolescents with specific language impairment. *Frontiers in Psychology, 5,* 175.

Hughes, C., Jaffee, S. R., Happé, F., Taylor, A., Caspi, A., and Moffitt, T. E. (2005). Origins of individual differences in theory of mind: From nature to nurture? *Child Development, 76*(2), 356–370.

Hull, D. L. (1986). On human nature. In *PSA: Proceedings of the biennial meeting of the philosophy of science association* (Vol. 1986, No. 2), Baltimore, MD: Philosophy of Science Association, 3–13.

Hurks, P. P. M. (2012). Does instruction in semantic clustering and switching enhance verbal fluency in children? *Clinical Neuropsychology, 26,* 1019–1037.

Insel, T. R., and Young, L. J. (2001). The neurobiology of attachment. *Nature Reviews Neuroscience, 2*(2), 129–136.

Ivanova, M. V., Isaev, D. Y., Dragoy, O. V., Akinina, Y. S., Petrushevskiy, A. G., Fedina, O. N., . . . and Dronkers, N. F. (2016). Diffusion-tensor imaging of major white matter tracts and their role in language processing in aphasia. *Cortex, 85,* 165–181.

Jablonka, E., and Lamb, M. J. (2005): *Evolution in Four Dimensions: Genetic, Epigenetic, Behavioral, and Symbolic Variation in the History of Life.* Cambridge, MA: MIT Press.

Jackson, P. L., Meltzoff, A. N., and Decety, J. (2006). Neural circuits involved in imitation and perspective-taking. *NeuroImage, 31*(1), 429–439.

Jacobson, S. W., and Kagan, J. (1979). Interpreting "imitative" responses in early infancy. *Science, 205,* 215–217.

James, W. (1890). *The Principles of Psychology.* New York: Holt.

Jelinek, E. (1995). Quantification in Straits Salish. In E. Bach, E. Jelinek, A. Kratzer, and B. Partee (eds.), *Quantification in Natural Languages.* Dordrecht, NL: Kluwer, 487–540.

Jensen, J., Smith, A. J., Willeit, M., Crawley, A. P., Mikulis, D. J., Vitcu, I., and Kapur, S. (2007). Separate brain regions code for salience versus valence during reward prediction in humans. *Human Brain Mapping, 28,* 294–302.

Johnson, M. H. (2005). Subcortical face processing. *Nature Reviews Neuroscience, 6*(10), 766–774.

Johnson, M. H., Dziurawiec, S., Ellis, H., and Morton, J. (1991). Newborns' preferential tracking of face-like stimuli and its subsequent decline. *Cognition, 40*(1), 1–19.

Johnson, S., Slaughter, V., and Carey, S. (1998). Whose gaze will infants follow? The elicitation of gaze-following in 12-month-olds. *Developmental Science, 1*(2), 233–238.

Jones, P. L., Ryan, M. J., Flores, V., and Page, R. A. (2013). When to approach novel prey cues? Social learning strategies in frog-eating bats. *Proceedings of the Royal Society of London, Series B: Biological Sciences, 280*(1772), 20132330.

Jones, S. S. (2006). Exploration or imitation? The effect of music on 4-week-old infants' tongue protrusions. *Infant Behavior and Development, 29*, 126–130.

Jones, S. S. (2007). Imitation in infancy: The development of mimicry. *Psychological Science, 18*, 593–599.

Jones, S. S. (2009). The development of imitation in infancy. *Philosophical Transactions of the Royal Society, Series B: Biological Sciences, 364*, 2325–2335.

Jusczyk, P. W. (1999). How infants begin to extract words from speech. *Trends in Cognitive Sciences, 3*(9), 323–328.

Kahneman, D. (2003). A perspective on judgment and choice: Mapping bounded rationality. *American Psychologist, 58*(9), 697–720.

Kano, F., Call, J., and Tomonaga, M. (2012). Face and eye scanning in gorillas (*Gorilla gorilla*), orangutans (*Pongo abelii*), and humans (*Homo sapiens*): Unique eye-viewing patterns in humans among hominids. *Journal of Comparative Psychology, 126*(4), 388–398.

Karmiloff-Smith, A. (1995). *Beyond Modularity: A Developmental Perspective on Cognitive Science*. Cambridge, MA: MIT Press.

Karmiloff-Smith, A. (2015). An alternative to domain-general or domain-specific frameworks for theorizing about human evolution and ontogenesis. *AIMS Neuroscience, 2*(2), 91–104.

Katz, J. J. (1985). An outline of Platonist grammar. In Katz, J. J (ed.), *The Philosophy of Linguistics*. Oxford: Oxford University Press.

Kaufman, S. B., DeYoung, C. G., Gray, J. R., Brown, J., and Mackintosh, N. (2009). Associative learning predicts intelligence above and beyond working memory and processing speed. *Intelligence, 37*, 374–382.

Kaye, K., and Marcus, J. (1978). Imitation over a series of trials without feedback: Age six months. *Infant Behavior and Development, 1*, 141–155.

Kepecs, A., and Mainen, Z. F. (2012). A computational framework for the study of confidence in humans and animals. *Philosophical Transactions of the Royal Society of London, Series B: Biological Sciences, 367*(1594), 1322–1337.

Keysers, C., Kohler, E., Umiltà, M. A., Nanetti, L., Fogassi, L., and Gallese, V. (2003). Audiovisual mirror neurons and action recognition. *Experimental Brain Research, 153*(4), 628–636.

Kidd, D. C., and Castano, E. (2013). Reading literary fiction improves theory of mind. *Science, 342*(6156), 377–380.

Kidd, E. (2012). Implicit statistical learning is directly associated with the acquisition of syntax. *Developmental Psychology, 48,* 171–184.

Kidd, E., and Arciuli, J. (2016). Individual differences in statistical learning predict children's comprehension of syntax. *Child Development, 87*(1), 184–193.

Kline, M. A., and Boyd, R. (2010). Population size predicts technological complexity in Oceania. *Proceedings of the Royal Society of London, Series B: Biological Sciences, 277*(1693), 2559–2564.

Klopfer, P. H. (1961). Observational learning in birds: The establishment of behavioral modes. *Behaviour, 17,* 71–80.

Koepke, J. E., Hamm, M., Legerstee, M., and Russell, M. (1983). Neonatal imitation: Two failures to replicate. *Infant Behavior and Development, 6,* 97–102.

Kohler, E., Keysers, C., Umiltà, M. A., Fogassi, L., Gallese, V., and Rizzolatti, G. (2002). Hearing sounds, understanding actions: Action representation in mirror neurons. *Science, 297*(5582), 846–848.

Kosfeld, M., Heinrichs, M., Zak, P. J., Fischbacher, U., and Fehr, E. (2005). Oxytocin increases trust in humans. *Nature, 435*(7042), 673–676.

Kovács, Á. M., Téglás, E., and Endress, A. D. (2010). The social sense: Susceptibility to others' beliefs in human infants and adults. *Science, 330*(6012), 1830–1834.

Kruglanski, A. W., and Gigerenzer, G. (2011). Intuitive and deliberate judgments are based on common principles. *Psychological Review, 118*(1), 97–109.

Krupenye, C., Kano, F., Hirata, S., Call, J., and Tomasello, M. (2016). Great apes anticipate that other individuals will act according to false beliefs. *Science, 354*(6308), 110–114.

Kuhl, P. K. (2007). Is speech learning "gated" by the social brain? *Developmental Science, 10*(1), 110–120.

Kuhn, T. S. (1962). *The Structure of Scientific Revolutions.* Chicago: University of Chicago Press.

Kuo, Z. Y. (1922). How are our instincts acquired? *Psychological Review, 29*(5), 344–365.

Lakoff, G. (1990). The invariance hypothesis: Is abstract reason based on image schemas? *Cognitive Linguistics, 1,* 39–74.

Laland, K. N. (2004). Social learning strategies. *Animal Learning & Behavior, 32*(1), 4–14.

Laland, K. N., and Brown, G. R. (2011). *Sense and Nonsense: Evolutionary Perspectives on Human Behaviour.* Oxford: Oxford University Press.

Laland, K. N., and Brown, G. R. (in press). The social construction of human nature. In T. Lewens and E. Hannon (eds.), *Why We Disagree about Human Nature.* Oxford: Oxford University Press.

Laland, K. N., and Rendell, L. (2013). Cultural memory. *Current Biology, 23,* R736–R740.

Lane, A., Mikolajczak, M., Treinen, E., Samson, D., Corneille, O., de Timary, P., and Luminet, O. (2015). Failed replication of oxytocin effects on trust: The envelope task case. *PloS One, 10*(9), e0137000.

Lawson, E. A., Marengi, D. A., DeSanti, R. L., Holmes, T. M., Schoenfeld, D. A., and Tolley, C. J. (2015). Oxytocin decreases caloric intake in men. *Obesity, 23,* 950–956.

Leadbeater, E. (2015). What evolves in the evolution of social learning? *Journal of Zoology, 295*(1), 4–11.

Lefebvre, L., and Giraldeau, L-A. (1996). Is social learning an adaptive specialization? In C. Heyes and B. G. Galef (eds.), *Social Learning in Animals: The Roots of Culture.* San Diego, CA: Academic Press, 107–152.

Leighton, J., Bird, G., and Heyes, C. (2010). "Goals" are not an integral component of imitation. *Cognition, 114*(3), 423–435.

Leighton, J., and Heyes, C. (2010). Hand to mouth: Automatic imitation across effector systems. *Journal of Experimental Psychology: Human Perception and Performance, 36*(5), 1174.

Lenneberg, E. H. (1967). *The Biological Foundations of Language.* New York: Wiley.

Lepage, J. F., and Théoret, H. (2007). The mirror neuron system: Grasping others' actions from birth? *Developmental Science, 10*(5), 513–523.

Le Pelley, M. E. (2010). Attention and human associative learning. In C. J. Mitchell and M. E. Le Pelley (eds.), *Attention and Associative Learning: From Brain to Behaviour.* New York: Oxford University Press, 187–215.

Le Pelley, M. E., Vadillo, M., and Luque, D. (2013). Learned predictiveness influences rapid attentional capture: Evidence from the dot probe task. *Journal of Experimental Psychology: Learning, Memory, and Cognition, 39*(6), 1888–1900.

Leslie, A. M. (1987). Pretense and representation: The origins of "theory of mind." *Psychological Review, 94*(4), 412–426.

Lewens, T. (2012). Human nature: The very idea. *Philosophy & Technology, 25*(4), 459–474.

Lewens, T. (2015). *Cultural Evolution: Conceptual Challenges.* Oxford: Oxford University Press.

Lewens, T. and Hannon, E. (eds.) (in press). *Why We Disagree about Human Nature.* Oxford: Oxford University Press.

Lewis, C., Freeman, N. H., Kyriakidou, C., Maridaki-Kassotaki, K., and Berridge, D. M. (1996). Social influences on false belief access: Specific sibling influences or general apprenticeship? *Child Development, 67*(6), 2930–2947.

Li, J. (2003). US and Chinese cultural beliefs about learning. *Journal of Educational Psychology, 95,* 258–267.

Li, W., Howard, J. D., Parrish, T. B., and Gottfried, J. A. (2008). Aversive learning enhances perceptual and cortical discrimination of indiscriminable odor cues. *Science, 319,* 1842–1845.

Lidz, J., and Gagliardi, A. (2015). How nature meets nurture: Universal grammar and statistical learning. *Annual Review of Linguistics, 1*(1), 333–353.

Lieberman, M. D. (2007). Social cognitive neuroscience: A review of core processes. *Annual Review of Psychology, 58,* 259–289.

Lillard, A. (1998). Ethnopsychologies: Cultural variations in theories of mind. *Psychological Bulletin, 123*(1), 3–32.

Lipton, P. (2003). *Inference to the Best Explanation.* London: Routledge.

Lohmann, H., and Tomasello, M. (2003). The role of language in the development of false belief understanding: A training study. *Child Development, 74*(4), 1130–1144.

Loman, M. M., and Gunnar, M. R. (2010). Early experience and the development of stress reactivity and regulation in children. *Neuroscience & Biobehavioral Reviews, 34*(6), 867–876.

Lorenz, K. (1965). *Evolution and the Modification of Behavior.* Chicago: University of Chicago Press.

Lorenz, R., and Tinbergen, N. (1938/1970). Taxis and instinctive behavior pattern in egg-rolling by the Graylag goose. In K. Lorenz (ed.), R. Martin (trans.), *Studies in Animal and in Human Behavior* (Vol. 1). Cambridge, MA: Harvard University Press, 316–359.

Lumsden, C. J., and Wilson, E. O. (2005). *Genes, Mind, and Culture: The Coevolutionary Process.* Singapore: World Scientific.

Luna, B. (2009). Developmental changes in cognitive control through adolescence. *Advances in Child Development & Behavior. 37,* 233–278.

Lunt, L., Bramham, J., Morris, R. G., Bullock, P. R., Selway, R. P., Xenitidis, K., and David, A. S. (2012). Prefrontal cortex dysfunction and "jumping to conclusions": Bias or deficit? *Journal of Neuropsychology, 6*(1), 65–78.

Lyons, D. E., Young, A. G., and Keil, F. C. (2007). The hidden structure of overimitation. *Proceedings of the National Academy of Sciences, 104*(50), 19751–19756.

Lyons, J. (1977). *Semantics.* London: Cambridge University Press.

Machery, E. (2008). A plea for human nature. *Philosophical Psychology, 21*(3), 321–329.

Machery, E. (in press). A plea for human nature, redux. In T. Lewens and E. Hannon (eds.), *Why We Disagree about Human Nature.* Oxford: Oxford University Press.

Mack, A., and Rock, I. (1998). *Inattentional Blindness.* Cambridge, MA: MIT Press.

Mackintosh, N. J. (1975). A theory of attention: Variations in the associability of stimuli with reinforcement. *Psychological Review, 82*(4), 276–298.

MacLean, E. L., Hare, B., Nunn, C. L., Addessi, E., Amici, F., Anderson, R. C., Aureli, F., Baker, J. M., Bania, A. E., Barnard, A. M., and Boogert, N. J. (2014). The evolution of self-control. *Proceedings of the National Academy of Sciences, 111*(20), E2140–E2148.

MacLeod, C., Mathews, A., and Tata, P. (1986). Attentional bias in emotional disorders. *Journal of Abnormal Psychology, 95*(1), 15–20.

MacPhail, E. M. (1982). *Brain and Intelligence in Vertebrates.* Oxford: Clarendon Press.

Mahmoodi, A., Bang, D., Ahmadabadi, M. N., and Bahrami, B. (2013). Learning to make collective decisions: The impact of confidence escalation. *PLoS One, 8*, e81195.

Margulis, E. H., Mlsna, L. M., Uppunda, A. K., Parrish, T. B., and Wong, P. C. (2009). Selective neurophysiologic responses to music in instrumentalists with different listening biographies. *Human Brain Mapping, 30*(1), 267–275.

Masters, J. C. (1979). Interpreting "imitative" responses in early infancy. *Science, 205*(4402), 215.

Matzel, L. D., and Kolata, S. (2010). Selective attention, working memory, and animal intelligence. *Neuroscience & Biobehavioral Reviews, 34*(1), 23–30.

Mayer, A., and Träuble, B. E. (2013). Synchrony in the onset of mental state understanding across cultures? A study among children in Samoa. *International Journal of Behavioral Development, 37*, 21–28.

Maynard-Smith, J. (2000). The concept of information in biology. *Philosophy of Science, 67*, 177–194.

McBrearty, S., and Brooks, A. S. (2000). The revolution that wasn't: A new interpretation of the origin of modern human behavior. *Journal of Human Evolution, 39*(5), 453–563.

McEwen, F., Happé, F., Bolton, P., Rijsdijk, F., Ronald, A., Dworzynski, K., and Plomin, R. (2007). Origins of individual differences in imitation: Links with language, pretend play, and socially insightful behavior in two-year-old twins. *Child Development, 78*(2), 474–492.

McGeer, V. (2007). The regulative dimension of folk psychology. In D. D. Hutto and M. Ratcliffe (eds.), *Folk Psychology Re-Assessed.* New York: Springer, 137–156.

McGuigan, N. (2013). The influence of model status on the tendency of young children to over-imitate. *Journal of Experimental Child Psychology, 116*(4), 962–969.

McKenzie, B., and Over, R. (1983). Young infants fail to imitate facial and manual gestures. *Infant Behavior and Development, 6,* 85–95.

Meins, E. (2012). Social relationships and children's understanding of mind: Attachment, internal states, and mind-mindedness. In M. Siegal and L. Surian (eds.), *Access to Language and Cognitive Development.* Oxford: Oxford University Press, 23–43.

Mellars, P. (1989). Major issues in the emergence of modern humans. *Current Anthropology, 30*(3), 348–385.

Mellars, P. (2005). The impossible coincidence. A single-species model for the origins of modern human behavior in Europe. *Evolutionary Anthropology, 14,* 12–27.

Mellars, P., and Stringer, C. (eds.) (1989). *The Human Revolution.* Edinburgh: Edinburgh University Press.

Meltzoff, A. N. (1988). The human infant as Homo imitans. In T. R. Zentall and B. G. Galef (eds.), *Social Learning: Psychological and Biological Perspectives* Hillsdale, NJ: Erlbaum, 319–341.

Meltzoff, A. N. (2002). Imitation as a mechanism of social cognition: Origins of empathy, theory of mind, and the representation of action. In U. Goswami (ed.), *Blackwell Handbook of Childhood Cognitive Development.* Oxford: Blackwell, 6–25.

Meltzoff, A. N. (2005). Imitation and other minds: The "like me" hypothesis. In S. Hurley and N. Chater (eds.), *Perspectives on Imitation: From Neuroscience to Social Science* (Vol. 2). Cambridge, MA: MIT Press, 55–78.

Meltzoff, A. N., and Moore, M. K. (1977). Imitation of facial and manual gestures by human neonates. *Science, 198*(4312), 75–78.

Meltzoff, A. N., and Moore, M. K. (1979). Interpreting "imitative" responses in early infancy. *Science, 205,* 217–219.

Meltzoff, A. N., and Moore, M. K. (1994). Imitation, memory, and the representation of persons. *Infant Behavior & Development, 17,* 83–99.

Meltzoff, A. N., and Moore, M. K. (1997). Explaining facial imitation: A theoretical model. *Early Development & Parenting, 6*(3–4), 179–192.

Mengotti, P., Ticini, L. F., Waszak, F., Schutz-Bosbach, S., and Rumiati, R. I. (2013). Imitating others' actions: Transcranial magnetic stimulation of the parietal opercula reveals the processes underlying automatic imitation. *European Journal of Neuroscience, 37*(2), 316–322.

Meristo, M., Hjelmquist, E., and Morgan, G. (2012). How access to language affects theory of mind in deaf children. In M. Siegal and L. Surian (eds.), *Access to Language and Cognitive Development.* New York: Oxford University Press, 44–61.

Mesoudi, A. (2011). *Cultural Evolution: How Darwinian Theory Can Explain Human Culture and Synthesize the Social Sciences.* Chicago: University of Chicago Press.

Mesoudi, A., Chang, L., Murray, K., and Jing Lu, H. (2015). Higher frequency of social learning in China than in the West shows cultural variation in the dynamics of cultural evolution. *Proceedings of the Royal Society of London, Series B: Biological Sciences,282,* 20142209.

Mesoudi, A., Whiten, A., and Laland, K. N. (2004). Perspective: Is human cultural evolution Darwinian? Evidence reviewed from the perspective of *The Origin of Species. Evolution, 58*(1), 1–11.

Michel, G. F., and Tyler, A. N. (2005). Critical period: A history of the transition from questions of when, to what, to how. *Developmental Psychobiology, 46*(3), 156–162.

Millikan, R. G. (1984). *Language, Thought, and Other Biological Categories: New Foundations for Realism.* Cambridge, MA: MIT press.

Misyak, J. B., and Christiansen, M. H. (2012). Statistical learning and language: An individual differences study. *Language Learning, 62*(1), 302–331.

Moerk, E. L. (1991). Positive evidence for negative evidence. *First Language, 11*(32), 219–251.

Molenberghs, P., Cunnington, R., and Mattingley, J. B. (2012). Brain regions with mirror properties: A meta-analysis of 125 human fMRI studies. *Neuroscience and Biobehavioral Reviews, 36*(1), 341–349.

Moon, C., Cooper, R. P., and Fifer, W. P. (1993). Two-day-olds prefer their native language. *Infant Behavior and Development, 16*(4), 495–500.

Moore, C., and Corkum, V. (1994). Social understanding at the end of the first year of life. *Developmental Review, 14*(4), 349–372.

Moore, J. W., Dickinson, A., and Fletcher, P. C. (2011). Sense of agency, associative learning, and schizotypy. *Consciousness and Cognition, 20,* 792–800.

Moore, R. (2016). Gricean communication and cognitive development. *The Philosophical Quarterly, 67*(267), 303–326.

Moore, R. (in press). Social cognition, stag hunts, and the evolution of language. *Biology & Philosophy.*

Morgan, T. J. H., Rendell, L. E., Ehn, M., Hoppitt, W., and Laland, K. N. (2012). The evolutionary basis of human social learning. *Proceedings of the Royal Society of London, Series B: Biological Sciences, 279*(1729), 653–662.

Morin, O. (2015). *How Traditions Live and Die.* Oxford: Oxford University Press.

Muthukrishna, M., and Henrich, J. (2016). Innovation in the collective brain. *Philosophical Transactions of the Royal Society, Series B, 371*(1690), 20150192.

Nicolle, A., Klein-Flügge, M. C., Hunt, L. T., Vlaev, I., Dolan, R. J., and Behrens, T. E. (2012). An agent independent axis for executed and modeled choice in medial prefrontal cortex. *Neuron, 75*(6), 1114–1121.

Nile, E., and Van Bergen, P. (2015). Not all semantics: Similarities and differences in reminiscing function and content between indigenous and non-indigenous Australians. *Memory, 23*(1), 83–98.

Nisbett, R. (2010). *The Geography of Thought: How Asians and Westerners Think Differently . . . and Why.* New York: Simon and Schuster.

Norman, D. A., and Shallice, T. (1986). Attention to action: Willed and automatic control of behaviour. In R. J. Davidson, et al. (eds.), *Consciousness and Self-regulation* (Vol. 4). Dordrecht, NL: Plenum, 1–18.

O'Brien, K., Slaughter, V., and Peterson, C. C. (2011). Sibling influences on theory of mind development for children with ASD. *Journal of Child Psychology and Psychiatry, 52*(6), 713–719.

Okasha, S. (2005). Multilevel selection and the major transitions in evolution. *Philosophy of Science, 72*(5), 1013–1025.

Olsson, A., and Phelps, E. A. (2007). Social learning of fear. *Nature Neuroscience, 10,* 1095–1102.

Onishi, K. H., and Baillargeon, R. (2005). Do 15-month-old infants understand false beliefs? *Science, 308*(5719), 255–258.

Oostenbroek, J., Suddendorf, T., Nielsen, M., Redshaw, J., Kennedy-Costantini, S., Davis, J., . . . and Slaughter, V. (2016). Comprehensive longitudinal study challenges the existence of neonatal imitation in humans. *Current Biology, 26*(10), 1334–1338.

Osada, T. (1992). *A Reference Grammar of Mundari.* Institute for the Study of Languages and Cultures of Asia and Africa, Tokyo University of Foreign Studies.

O'Toole, R., and Dubin, R. (1968). Baby feeding and body sway: An experiment in George Herbert Mead's "Taking the role of the other." *Journal of Personality and Social Psychology, 10,* 59–65.

Ozturk, O., Krehm, M., and Vouloumanos, A. (2013). Sound symbolism in infancy: Evidence for sound-shape cross-modal correspondences in 4-month-olds. *Journal of Experimental Child Psychology, 114*(2), 173–186.

Pagel, M. (2000). The history, rate and pattern of world linguistic evolution. In C. Knight, M. Studdert-Kennedy, and J. Hurford (eds.), *The Evolutionary Emergence of Language.* Cambridge: Cambridge University Press, 391–416.

Papineau, D. (1987). *Reality and Representation.* Oxford: Basil Blackwell.

Paracchini, S., Scerri, T., and Monaco, A. P. (2007). The genetic lexicon of dyslexia. *Annual Review of Genomics, Human Genetics, 8,* 57–79.

Pascalis, O., de Haan, M., and Nelson, C. A. (2002). Is face processing species-specific during the first year of life? *Science, 296*(5571), 1321–1323.

Passingham, R. E. (2008). *What is Special about the Human Brain?* New York: Oxford University Press.

Passingham, R. E., and Smaers, J. B. (2014). Is the prefrontal cortex especially enlarged in the human brain? Allometric relations and remapping factors. *Brain, Behavior and Evolution, 84*(2), 156–166.

Paulus, M., Hunnius, S., Vissers, M., and Bekkering, H. (2011). Imitation in infancy: Rational or motor resonance? *Child Development, 82*(4), 1047–1057.

Pawlby, S. J. (1977). Imitative interaction. In H. Schaffer (ed.), *Studies in Mother–Infant Interaction.* New York: Academic Press, 203–224.

Pearce, J. M. (2013). *Animal Learning and Cognition: An Introduction.* London: Psychology Press.

Perner, J. (2010). Who took the cog out of cognitive science? In P. A. Frensch and R. Schwarzer (eds.), *Cognition and Neuropsychology: International Perspectives on Psychological Science* (Vol. 1). London: Psychology Press, 241–261.

Petersen, S. E., and Posner, M. I. (2012). The attention system of the human brain 20 years after. *Annual Review of Neuroscience, 35,* 73–89.

Piaget, J. (1962). *Play, Dreams and Imitation in Childhood.* New York: Norton.

Pinker, S. (1994). *The Language Instinct: The New Science of Language and Mind.* London: Penguin.

Pinker, S. (1997). *How the Mind Works.* London: Penguin.

Pinker, S., and Bloom, P. (1990). Natural language and natural selection. *Behavioral and Brain Sciences, 13*(4), 707–727.

Pinker, S., and Jackendoff, R. (2005). The faculty of language: What's special about it? *Cognition, 95*(2), 201–236.

Plomin, R., DeFries, J. C., McClearn, G. E., and McGuffin, P. (2001). *Behavior Genetics.* New York: W. H. Freeman.

Plotkin, H. C., and Odling-Smee, F. J. (1981). A multiple-level model of evolution and its implications for sociobiology. *Behavioral and Brain Sciences, 4*(2), 225–235.

Pocheville, A., and Danchin, E. (2017). Genetic assimilation and the paradox of blind variation. In P. Huneman and D. Walsh (eds.), *Challenging the Modern Synthesis.* Oxford: Oxford University Press.

Poldrack, R. A. (2006). Can cognitive processes be inferred from neuroimaging data? *Trends in Cognitive Sciences, 10*(2), 59–63.

Popper, K. (1962). *Conjectures and Refutations.* New York: Basic Books.

Prados, J. (2011). Blocking and overshadowing in human geometry learning. *Journal of Experimental Psychology: Animal Behavior Processes, 37*, 121–126.

Press, C., Bird, G., Flach, R., and Heyes, C. (2005). Robotic movement elicits automatic imitation. *Cognitive Brain Research, 25*(3), 632–640.

Pyers, J. E., and Senghas, A. (2009). Language promotes false-belief understanding: Evidence from learners of a new sign language. *Psychological Science, 20*(7), 805–812.

Quiroga, R. Q., Reddy, L., Kreiman, G., Koch, C., and Fried, I. (2005). Invariant visual representation by single neurons in the human brain. *Nature, 435*(7045), 1102–1110.

Qureshi, A. W., Apperly, I. A., and Samson, D. (2010). Executive function is necessary for perspective selection, not Level-1 visual perspective calculation: Evidence from a dual-task study of adults. *Cognition, 117*(2), 230–236.

Rankin, C. H., Abrams, T., Barry, R. J., Bhatnagar, S., Clayton, D. F., Colombo, J., Coppola, G., Geyer, M. A., Glanzman, D. L., Marsland, S., and McSweeney, F. K. (2009). Habituation revisited: An updated and revised description of the behavioral characteristics of habituation. *Neurobiology of Learning and Memory, 92*(2), 135–138.

Rashotte, M. E., Griffin, R. W., and Sisk, C. L. (1977). Second-order conditioning of the pigeon's keypeck. *Learning & Behavior, 5*(1), 25–38.

Ray, E., and Heyes, C. (2011). Imitation in infancy: The wealth of the stimulus. *Developmental Science, 14*(1), 92–105.

Reader, S. M., Hager, Y., and Laland, K. N. (2011). The evolution of primate general and cultural intelligence. *Philosophical Transactions of the Royal Society, Series B, 366,* 1017–1027.

Reader, S. M., and Laland, K. N. (2002). Social intelligence, innovation, and enhanced brain size in primates. *Proceedings of the National Academy of Sciences, 99,* 4436–4441.

Reeb-Sutherland, B. C., Fifer, W. P., Byrd, D. L., Hammock, E. A., Levitt, P., and Fox, N. A. (2011). One-month-old human infants learn about the social world while they sleep. *Developmental Science, 14*(5), 1134–1141.

Reeb-Sutherland, B. C., Levitt, P., and Fox, N. A. (2012). The predictive nature of individual differences in early associative learning and emerging social behavior. *PloS One, 7*(1), e30511.

Reid, V. M., Dunn, K., Young, R. J., Amu, J., Donovan, T., and Reissland, N. (2017). The human fetus preferentially engages with face-like visual stimuli. *Current Biology, 27,* 1825–1828.

Reimers-Kipping, S., Hevers, W., Pääbo, S., and Enard, W. (2011). Humanized FOXP2 specifically affects cortico-basal ganglia circuits. *Neuroscience, 175,* 75–84.

Rendell, L., Fogarty, L., Hoppitt, W. J., Morgan, T. J., Webster, M. M., and Laland, K. N. (2011). Cognitive culture: Theoretical and empirical insights into social learning strategies. *Trends in Cognitive Sciences, 15*(2), 68–76.

Rescorla, R. A. (1988). Pavlovian conditioning: It's not what you think it is. *American Psychologist, 43*(3), 151–160.

Rescorla, R. A., and Wagner, A. R. (1972). A theory of Pavlovian conditioning: Variations in the effectiveness of reinforcement and nonreinforcement. *Classical Conditioning II: Current Research and Theory, 2,* 64–99.

Richerson, P. J. (in press). What work (or mischief) does "human nature" do in the work of scientists? In T. Lewens and E. Hannon (eds.), *Why We Disagree about Human Nature.* Oxford: Oxford University Press.

Richerson, P. J., and Boyd, R. (2005). *Not by Genes Alone.* Chicago: University of Chicago Press.

Richerson, P. J., and Boyd, R. (2013). Rethinking paleoanthropology: A world queerer than we supposed. In G. Hatfield and H. Pittman (eds.), *Evolution of Mind, Brain, and Culture.* Philadelphia: University of Pennsylvania Press, 263–302.

Rilling, J. K. (2014). Comparative primate neuroimaging: Insights into human brain evolution. *Trends in Cognitive Sciences, 18*(1), 46–55.

Rizzolatti, G., and Craighero, L. (2004). The mirror-neuron system. *Annual Review of Neuroscience, 27,* 169–192.

Rogers, A. R. (1988). Does biology constrain culture? *American Anthropologist, 90,* 819–831.

Rosa-Salva, O., Regolin, L., and Vallortigara, G. (2010). Faces are special for newly hatched chicks: Evidence for inborn domain-specific mechanisms underlying spontaneous preferences for face-like stimuli. *Developmental Science, 13*(4), 565–577.

Rovee-Collier, C., and Giles, A. (2010). Why a neuromaturational model of memory fails: Exuberant learning in early infancy. *Behavioural Processes, 83*(2), 197–206.

Rovee-Collier, C., Mitchell, K., and Hsu-Yang, V. (2013). Effortlessly strengthening infant memory: Associative potentiation of new learning. *Scandinavian Journal of Psychology, 54*(1), 4–9.

Rushworth, M. F., Mars, R. B., and Sallet, J. (2013). Are there specialized circuits for social cognition, and are they unique to humans? *Current Opinion in Neurobiology, 23*(3), 436–442.

Ryle, G. (1945). "Philosophical Arguments." Inaugural lecture, Waynflete Professor of Metaphysical Philosophy. Reprinted (2009) in *Collected Papers of Gilbert Ryle,* Vol. 2. London: Routledge, 194–211.

Saggerson, A. L., George, D. N., and Honey, R. C. (2005). Imitative learning of stimulus-response and response-outcome associations in pigeons. *Journal of Experimental Psychology: Animal Behavior Processes, 31,* 289–300.

Salmon, K., and Reese, E. (2016). The benefits of reminiscing with young children. *Current Directions in Psychological Science, 25*(4), 233–238.

Samson, D., Apperly, I. A., Braithwaite, J. J., Andrews, B. J., and Bodley Scott, S. E. (2010). Seeing it their way: Evidence for rapid and involuntary computation of what other people see. *Journal of Experimental Psychology: Human Perception and Performance, 36*(5), 1255.

Samuels, R. (2004). Innateness in cognitive science. *Trends in Cognitive Sciences, 8*(3), 136–141.

Samuels, R. (2012). Science and human nature. *Royal Institute of Philosophy Supplement, 70,* 1–28.

Santiesteban, I., Catmur, C., Hopkins, S. C., Bird, G., and Heyes, C. (2014). Avatars and arrows: Implicit mentalizing or domain-general processing? *Journal of Experimental Psychology: Human Perception and Performance, 40*(3), 929–937.

Scheele, D., Wille, A., Kendrick, K. M., Stoffel-Wagner, B., Becker, B., Gunturkun, O., Maier, W., and Hurlemann, R. (2013). Oxytocin enhances brain reward system responses of men viewing the face of their female partner. *Proceedings of the National Academy of Sciences, 110,* 20308–20313.

Schreiweis, C., Bornschein, U., Burguière, E., Kerimoglu, C., Schreiter, S., Dannemann, M., Goyal, S., Rea, E., French, C. A., Puliyadi, R., and Groszer, M. (2014). Humanized FOXP2 accelerates learning by enhancing transitions from declarative to procedural performance. *Proceedings of the National Academy of Sciences, 111*(39), 14253–14258.

Schultz, W., and Dickinson, A. (2000). Neuronal coding of prediction errors. *Annual Review of Neuroscience, 23*(1), 473–500.

Senju, A., and Csibra, G. (2008). Gaze following in human infants depends on communicative signals. *Current Biology, 18*(9), 668–671.

Senju, A., Southgate, V., White, S., and Frith, U. (2009). Mindblind eyes: An absence of spontaneous theory of mind in Asperger syndrome. *Science, 325*(5942), 883–885.

Senzaki, S., Masuda, T., and Ishii, K. (2014). When is perception top-down and when is it not? Culture, narrative, and attention. *Cognitive Science, 38*(7), 1493–1506.

Shahaeian, A., Peterson, C. C., Slaughter, V., and Wellman, H. M. (2011). Culture and the sequence of steps in theory of mind development. *Developmental Psychology, 47*(5), 1239–1247.

Shanks, D. R. (2010). Learning: from association to cognition. *Annual Review of Psychology, 61,* 273–301.

Shannon, C. E. (1949). The mathematical theory of communication. In C. E. Shannon and W. Weaver (eds.), *The Mathematical Theory of Communication.* Urbana: University of Illinois Press.

Shea, N. (2009). Imitation as an inheritance system. *Philosophical Transactions of the Royal Society of London, Series B: Biological Sciences, 364,* 2429–2443.

Shea, N. (2013). Inherited representations are read in development. *The British Journal for the Philosophy of Science, 64*(1), 1–31.

Shea, N., Boldt, A., Bang, D., Yeung, N., Heyes, C., and Frith, C. D. (2014). Suprapersonal cognitive control and metacognition. *Trends in Cognitive Sciences, 18*(4), 186–193.

Shettleworth, S. J. (2010). *Cognition, Evolution, and Behavior.* Oxford: Oxford University Press.

Shiraev, E., and Levy, D. A. (2014). *Cross-Cultural Psychology.* New York: Pearson Education Limited.

Shweder, R. A., and Sullivan, M. A. (1993). Cultural psychology: Who needs it? *Annual Review of Psychology, 44*(1), 497–523.

Silk, J. B., and House, B. R. (2011). Evolutionary foundations of human prosocial sentiments. *Proceedings of the National Academy of Sciences, 108*(2), 10910–10917.

Simons, D. J., and Chabris, C. F. (1999). Gorillas in our midst: Sustained inattentional blindness for dynamic events. *Perception, 28*(9), 1059–1074.

Siqueland, E. R., and Lipsitt, L. P. (1966). Conditioned head-turning in human newborns. *Journal of Experimental Child Psychology, 3*(4), 356–376.

Skerry, A. E., Carey, S. E., and Spelke, E. S. (2013). First-person action experience reveals sensitivity to action efficiency in prereaching infants. *Proceedings of the National Academy of Sciences, 110*(46), 18728–18733.

Skuse, D. H. (1993). Extreme deprivation in early childhood. In D. Bishop and K. Mogford (eds.), *Language Development in Exceptional Circumstances.* Hillsdale, NJ: Erlbaum, 29–46.

Skuse, D. H., and Gallagher, L. (2011). Genetic influences on social cognition. *Pediatric Research, 69,* 85R–91R.

Slaughter, V., and Peterson, C. C. (2012). How conversational input shapes theory of mind development in infancy and early childhood. In M. Siegal and L. Surian (eds.), *Access to Language and Cognitive Development.* Oxford: Oxford University Press, 3–22.

Smith, L. B., Suanda, S. H., and Yu, C. (2014). The unrealized promise of infant statistical word–referent learning. *Trends in Cognitive Sciences, 18*(5), 251–258.

Smith, N. A., and Trainor, L. J. (2008). Infant-directed speech is modulated by infant feedback. *Infancy, 13*(4), 410–420.

Snowdon, C. T. (2004). Social processes in the evolution of complex cognition and communication. In D. Kimbrough Oller and U. Griebel (eds.), *Evolution of Communication Systems: A Comparative Approach.* Cambridge, MA: MIT Press, 131–150.

Sober, E. (1991). Models of cultural evolution. In P. Griffiths (ed.), *Trees of Life: Essays in the Philosophy of Biology.* Dordrecht, NL: Kluwer Academic Publishers.

Southgate, V., Senju, A., and Csibra, G. (2007). Action anticipation through attribution of false belief by 2-year-olds. *Psychological Science, 18*(7), 587–592.

Spelke, E. S., and Kinzler, K. D. (2007). Core knowledge. *Developmental Science, 10*(1), 89–96.

Sperber, D. (1996). *Explaining Culture.* Oxford: Blackwell Publishers.

Sperber, D. (2000a). An objection to the memetic approach to culture. In R. Aunger (ed.), *Darwinizing Culture: The Status of Memetics as a Science.* Oxford: Oxford University Press, 163–174.

Sperber, D. (2000b). Metarepresentations in an evolutionary perspective. In D. Sperber (ed.), *Metarepresentations: A Multidisciplinary Perspective.* Oxford: Oxford University Press, 117–137.

Sperber, D., and Wilson, D. (1995). *Relevance: Communication and Cognition.* 2nd ed. Cambridge: Cambridge University Press.

Sterelny, K. (2003). *Thought in a Hostile World: The Evolution of Human Cognition.* Oxford: Wiley-Blackwell.

Sterelny, K. (2006). Memes revisited. *The British Journal for the Philosophy of Science, 57*(1), 145–165.

Sterelny, K. (2012). *The Evolved Apprentice.* Cambridge, MA: MIT Press.

Sterelny, K. (2017). Cultural evolution in California and Paris. *Studies in the History and Philosophy of Biological and Biomedical Sciences, 62,* 42–50.

Sterelny, K. (in press). Culture and the extended phenotype: Cognition and material culture in deep time. In A. Newen, L. de Bruin, and S. Gallagher (eds.), *The Oxford Handbook of Cognition: Embodied, Embedded, Enactive and Extended.* Oxford: Oxford University Press.

Sterelny, K., Smith, K. C., and Dickison, M. (1996). The extended replicator. *Biology and Philosophy, 11*(3), 377–403.

Street, J. A., and Dąbrowska, E. (2010). More individual differences in language attainment: How much do adult native speakers of English know about passives and quantifiers? *Lingua, 120*(8), 2080–2094.

Tamariz, M., and Kirby, S. (2016). The cultural evolution of language. *Current Opinion in Psychology, 8,* 37–43.

Tamir, M., and Robinson, M. D. (2007). The happy spotlight: Positive mood and selective attention to rewarding information. *Personality and Social Psychology Bulletin, 33*(8), 1124–1136.

Tarr, B., Launay, J., Cohen, E., and Dunbar, R. (2015). Synchrony and exertion during dance independently raise pain threshold and encourage social bonding. *Biology Letters, 11*(10), 20150767.

Taumoepeau, M. (2016). Maternal expansions of child language relate to growth in children's vocabulary. *Language Learning and Development, 12*(4), 429–446.

Taumoepeau, M., and Ruffman, T. (2006). Mother and infant talk about mental states relates to desire language and emotion understanding. *Child Development, 77*(2), 465–481.

Taumoepeau, M., and Ruffman, T. (2008). Stepping stones to others' minds: Maternal talk relates to child mental state language and emotion understanding at 15, 24, and 33 months. *Child Development, 79,* 284–302.

Templeton, J. J., Kamil, A. C., and Balda, R. P. (1999). Sociality and social learning in two species of corvids: The Pinyon Jay and the Clark's Nutcracker. *Journal of Comparative Psychology, 113,* 450–455.

Ten Cate, C., and Okanoya, K. (2012). Revisiting the syntactic abilities of non-human animals: Natural vocalizations and artificial grammar learning. *Philosophical Transactions of the Royal Society of London, Series B: Biological Sciences, 367*(1598), 1984–1994.

Tennie, C., Call, J., and Tomasello, M. (2009). Ratcheting up the ratchet: On the evolution of cumulative culture. *Philosophical Transactions of the Royal Society, Series B: Biological Sciences, 364,* 2405–2415.

Thornton, A., and McAuliffe, K. (2012). Teaching can teach us a lot. *Animal Behaviour, 83*(4), e6–e9.

Thornton, A., and Raihani, N. J. (2008). The evolution of teaching. *Animal Behaviour, 75*(6), 1823–1836.

Thurman, S. M., and Lu, H. (2013). Physical and biological constraints govern perceived animacy of scrambled human forms. *Psychological Science, 24,* 1133–1141.

Tinbergen, N. (1963). On aims and methods of ethology. *Zeitschrift für Tierpsychologie, 20*(4), 410–433.

Tipples, J. (2008). Orienting to counterpredictive gaze and arrow cues. *Perception & Psychophysics, 70*(1), 77–87.

Toelch, U., Bruce, M., Newson, L., Richerson, P. J., and Reader, S. M. (2014). Individual consistency and flexibility in human social information use. *Proceedings of the Royal Society of London, Series B: Biological Sciences, 281,* 20132864.

Tomasello, M. (1995). Language is not an instinct. *Cognitive Development, 10,* 131–156.

Tomasello, M. (2003). *Constructing a Language. A Usage-Based Approach.* Cambridge, MA: Harvard University Press.

Tomasello, M. (2009). *The Cultural Origins of Human Cognition.* Cambridge, MA: Harvard University Press.

Tomasello, M. (2014). *A Natural History of Human Thinking.* Cambridge, MA: Harvard University Press.

Tomasello, M., Carpenter, M., Call, J., Behne, T., and Moll, H. (2005). Understanding and sharing intentions: The origins of cultural cognition. *Behavioral and Brain Sciences, 28,* 675–735.

Tomasello M., Kruger A. C., and Ratner H. H. (1993). Cultural learning. *Behavioral & Brain Sciences, 16,* 495–552.

Tomasello, M., and Moll, H. (2010). The gap is social: Human shared intentionality and culture. In P. Kapperler and J. Silk (eds.), *Mind the Gap. Tracing the Origins of Human Universals.* Berlin: Springer, 331–349.

Tomblin, J. B., Mainela-Arnold, E., and Zhang, X. (2007). Procedural learning in adolescents with and without specific language impairment. *Language Learning and Development, 3*(4), 269–293.

Tomblin, J. B., Shriberg, L., Murray, J., Patil, S., and Williams, C. (2004). Speech and language characteristics associated with a 7 / 13 translocation involving FOXP2. *American Journal of Medical Genetics, Part B, Neuropsychiatric Genetics, 130,* 97–97.

Trainor, L. J., Austin, C. M., and Desjardins, R. N. (2000). Is infant-directed speech prosody a result of the vocal expression of emotion? *Psychological Science, 11*(3), 188–195.

Trainor, L. J., and Desjardins, R. N. (2002). Pitch characteristics of infant-directed speech affect infants' ability to discriminate vowels. *Psychonomic Bulletin & Review, 9*(2), 335–340.

Triesch, J., Teuscher, C., Deák, G. O., and Carlson, E. (2006). Gaze following: Why (not) learn it? *Developmental Science, 9*(2), 125–147.

Tunçgenç, B., and Cohen, E. (2016). Movement synchrony forges social bonds across group divides. *Frontiers in Psychology, 7,* 782.

Uller, T. (2008). Developmental plasticity and the evolution of parental effects. *Trends in Ecology and Evolution, 23,* 432–438.

Uzgiris, I. C. (1972). Patterns of vocal and gestural imitation in infants. In F. Monks, W. Hartup, and J. de Witt (eds.), *Determinants of Behavioral Development.* New York: Academic Press, 467–471.

Uzgiris, I. C., Benson, J. B., Kruper, J. C., and Vasek, M. E. (1989). Contextual influences on imitative interactions between mothers and infants. In J. Lockman and N. Hazen (eds.), *Action in Social Context: Perspectives on Early Development.* New York: Plenum Press, 103–127.

Van de Waal, E., Renevey, N., Farve, C. M., and Bahary, R. (2010). Selective attention to philopatric models causes directed social learning in wild vervet monkeys. *Proceedings of the Royal Society of London, Series B: Biological Sciences, 282,* 20092260.

Van der Lely, H. K., and Pinker, S. (2014). The biological basis of language: Insight from developmental grammatical impairments. *Trends in Cognitive Sciences, 18*(11), 586–595.

Van Overwalle, F. (2009). Social cognition and the brain: A meta-analysis. *Human Brain Mapping, 30*(3), 829–858.

Van Schaik, C. P., and Pradhan, G. R. (2003). A model for tool-use traditions in primates: Implications for the coevolution of culture and cognition. *Journal of Human Evolution, 44*, 645–664.

Vecera, S. P., and Johnson, M. H. (1995). Gaze detection and the cortical processing of faces: Evidence from infants and adults. *Visual Cognition, 2*(1), 59–87.

Vetter, N. C., Leipold, K., Kliegel, M., Phillips, L. H., and Altgassen, M. (2013). Ongoing development of social cognition in adolescence. *Child Neuropsychology, 19*(6), 615–629.

Voelkl, B., and Huber, L. (2007). Imitation as faithful copying of a novel technique in marmoset monkeys. *PLoS One, 2*(7), e611.

Vogt, S., Buccino, G., Wohlschlager, A. M., Canessa, N., Shah, N. J., Zilles, K., Eickhoff, S. B., Freund, H. J., Rizzolatti, G., and Fink, G. R. (2007). Prefrontal involvement in imitation learning of hand actions: Effects of practice and expertise. *NeuroImage, 37*(4), 1371–1383.

Vouloumanos, A., Hauser, M. D., Werker, J. F., and Martin, A. (2010). The tuning of human neonates' preference for speech. *Child Development, 81*(2), 517–527.

Vouloumanos, A., and Werker, J. F. (2007). Listening to language at birth: Evidence for a bias for speech in neonates. *Developmental Science, 10*(2), 159–164.

Washburn, M. F. (1908). *The Animal Mind.* New York: Macmillan.

Watson, J. B. (1930). *Behaviorism.* Chicago: Phoenix.

Watson, J. S., and Ramey, C. T. (1972). Reactions to response-contingent stimulation in early infancy. *Merrill-Palmer Quarterly of Behavior and Development, 18*(3), 219–227.

Webster, M. M., and Laland, K. N. (2015). Space-use and sociability are not related to public-information use in ninespine sticklebacks. *Behavioral Ecology and Sociobiology, 69*(6), 895–907.

Werker, J. F., and Hensch, T. K. (2015). Critical periods in speech perception: New directions. *Annual Review of Psychology, 66*, 173–196.

Werner, C., and Latane, B. (1974). Interaction motivates attraction: Rats are fond of fondling. *Journal of Personality and Social Psychology, 29*(3), 328–334.

West-Eberhard, M. J. (2003). *Developmental Plasticity and Evolution.* Oxford: Oxford University Press.

West-Eberhard, M. J. (2005). Developmental plasticity and the origin of species differences. *Proceedings of the National Academy of Sciences, 102*(1), 6543–6549.

White, B. L., Castle, P., and Held, R. (1964). Observations on the development of visually guided reaching. *Child Development, 35,* 349–364.

Whitehouse, H. (2004). *Modes of Religiosity: A Cognitive Theory of Religious Transmission.* Lanham, MA: Rowman Altamira.

Whiten, A., and Custance, D. (1996). Studies of imitation in chimpanzees and children. In C. Heyes and B. G. Galef (eds.), *Social Learning in Animals: The Roots of Culture.* San Diego: Academic Press, 291–318.

Whiten, A., Goodall, J., McGrew, W. C., Nishida, T., Reynolds, V., Sugiyama, Y., Tutin, C. E., Wrangham, R. W., and Boesch, C. (1999). Cultures in chimpanzees. *Nature, 399,* 682–685.

Whiten, A., and Ham, R. (1992). On the nature and evolution of imitation in the animal kingdom: Reappraisal of a century of research. *Advances in the Study of Behaviour, 21,* 239–283.

Whiten, A., McGuigan, N., Marshall-Pescini, S., and Hopper, L. M. (2009). Emulation, imitation, over-imitation and the scope of culture for child and chimpanzee. *Philosophical Transactions of the Royal Society of London, Series B: Biological Sciences, 364*(1528), 2417–2428.

Wilkins, A. S., Wrangham, R. W., and Fitch, W. T. (2014). The "domestication syndrome" in mammals: A unified explanation based on neural crest cell behavior and genetics. *Genetics, 197*(3), 795–808.

Wilkinson, A., Kuenstner, K., Mueller, J., and Huber, L. (2010). Social learning in a non-social reptile. *Biology Letters, 6,* 614–616.

Williamson, R. A., Meltzoff, A. N., and Markman, E. M. (2008). Prior experiences and perceived efficacy influence 3-year olds' imitation. *Developmental Psychology, 44*(1), 275–285.

Wilson, B., Slater, H., Kikuchi, Y., Milne, A. E., Marslen-Wilson, W. D., Smith, K., and Petkov, C. I. (2013). Auditory artificial grammar learning in macaque and marmoset monkeys. *Journal of Neuroscience, 33*(48), 18825–18835.

Wilson, E. O. (1975). *Sociobiology: The New Synthesis.* Cambridge, MA: Harvard University Press.

Wiltermuth, S. S., and Heath, C. (2009). Synchrony and cooperation. *Psychological Science, 20*(1), 1–5.

Winawer, J., Witthoft, N., Frank, M. C., Wu, L., Wade, A. R., and Boroditsky, L. (2007). Russian blues reveal effects of language on color discrimination. *Proceedings of the National Academy of Sciences, 104*(19), 7780–7785.

Wisdom, T. N., Song, X., and Goldstone, R. L. (2013). Social learning strategies in networked groups. *Cognitive Science* 37, 1383–1425.

Wobber, V., Hare, B., Lipson, S., Wrangham, R., and Ellison, P. (2013). Different ontogenetic patterns of testosterone production reflect divergent male reproductive strategies in chimpanzees and bonobos. *Physiology & Behavior, 116,* 44–53.

Woodward, J. (2010). Causation in biology: Stability, specificity, and the choice of levels of explanation. *Biology & Philosophy, 25*(3), 287–318.

Wunderlich, K., Symmonds, M., Bossaerts, P., and Dolan, R. J. (2011). Hedging your bets by learning reward correlations in the human brain. *Neuron, 71,* 1141–1152.

Yantis, S., and Jonides, J. (1990). Abrupt visual onsets and selective attention: Voluntary versus automatic allocation. *Journal of Experimental Psychology: Human Perception and Performance, 16*(1), 121–134.

Yong, E. (2012). Dark side of love. *New Scientist, 213,* 39–41.

Young, J. M., Krantz, P. J., McClannahan, L. E., and Poulson, C. L. (1994). Generalized imitation and response-class formation in children with autism. *Journal of Applied Behavior Analysis, 27*(4), 685–697.

Zentall, T. R. (2006). Imitation: Definitions, evidence, and mechanisms. *Animal Cognition, 9*(4), 335–353.

Zilles, K. (2005). Evolution of the human brain and comparative cyto- and receptor architecture. In S. Dehaene, J. R. Duhamel, M. D. Hauser, and G. Rizzolatti (eds.), *From Monkey Brain to Human Brain.* Cambridge, MA: MIT Press, 41–56.

Zuhurudeen, F. M., and Huang, Y. T. (2016). Effects of statistical learning on the acquisition of grammatical categories through Qur'anic memorization: A natural experiment. *Cognition, 148,* 79–84.

Notes

INTRODUCTION

1. The Oxford English Dictionary defines a gadget as "a comparatively small fitting, contrivance, or piece of mechanism." I have chosen the term primarily because it sounds like instinct (two spikey syllables) but means something very different. While instincts are in the genes, gadgets are "contrivances"—assembled through human action. I also embrace the implication that cognitive gadgets are "comparatively small fittings"; they work in close and continuous interaction with phylogenetically ancient cognitive mechanisms, and it is the latter—the mechanisms we share with other animals—that do the lion's share of controlling behavior. However, I am not suggesting that cognitive gadgets are products of intelligent design (that people intend to produce them), or that they are "small" in the sense of being unimportant. Distinctively human cognitive mechanisms are powerful and play crucial roles in enabling us to lead our peculiar lives.

2. NATURE, NURTURE, CULTURE

1. The teleosemantic view is not alone in offering an account of information with the potential to distinguish more and less important contributions to development (Griffiths et al., 2015; Woodward, 2010). Therefore, the

plausibility of cultural evolutionary psychology does not depend specifically on the strength of teleosemantics.

2. Some versions of teleosemantics would allow that Jennifer Aniston neurons represent Jennifer Aniston if, for example, they acquire their response properties through a learning mechanism that was favored by natural selection because it supported face recognition (Millikan, 1984), or if the learning mechanism is a selection process in its own right (Papineau, 1987). These refinements complicate the Jennifer Aniston story but preserve the basic principle of teleosemantics: representational content depends on selection.
3. It is likely that calluses became encoded in the ostrich genome through "genetic assimilation;" the early ancestors of contemporary ostriches developed calluses through abrasion alone. The calluses were functional—they protected the body from harm. Therefore, there was selection in favor of genetic mutations that progressively increased the rate at which abrasion produced calluses. Members of the population with these "accelerator" genes out-reproduced those who lacked the genes until no abrasion at all was necessary for the development of calluses. Genetic assimilation does not imply "Lamarckian inheritance" (Pocheville and Danchin, 2017).
4. In recent commentaries, philosophers of biology have raised questions about whether the California school is really offering selectionist analyses of cultural evolution, but there can be no doubt that the Californians often use the language of selection (Heyes, in press, a; Lewens, 2015; Sterelny, 2017).
5. Cultural evolutionists of the California school sometimes talk about the evolution of "ideas" or "beliefs." However, their models typically bind each "belief" to a single behavior—e.g., the "belief" that it is good to have a small family is ascribed to all, and only, those people who have small families—making them functionally equivalent to models of behavior change.

3. STARTER KIT

1. Many thanks to Uta Frith for introducing me to this term.
2. This picture of cognitive development is like Carey's (2009) and Karmiloff-Smith's (1995) in suggesting discontinuous change. New ways of thinking are constructed in the course of development through social interaction. However, because I am concerned with the human capacity for cultural evolution, my picture is more tightly focused on the development of social

cognition, and my "starter kit" is much leaner than the "core knowledge" posited by Carey, Spelke (Spelke and Kinzler, 2007), and Baillargeon (Baillargeon, Scott, and He, 2010). For example, I doubt that we genetically inherit an "agent" concept, and I have more confidence in associative learning as one of the engines of cognitive development. These differences arise from, and are expressed in, our differing interpretations of studies purporting to show that very young infants can, for example, imitate actions (Chapter 6) and represent false beliefs (Chapter 7).

3. Carey (2009) argues that gaze-cuing could not be due to a "learned association between gaze shift and something interesting in the direction of gaze" because infants are more likely to turn in the same direction as a faceless robot—an object that, lacking a face, does not have a "gaze"—provided that the robot has recently responded contingently to the infant's actions (Johnson, Slaughter, and Carey, 1998). However, this interesting result can also be interpreted as: (1) confirming that not all motion cuing is gaze-cuing, and (2) showing that motion cues are more effective when they come from an object that the infant likes, and possibly attends to more closely, because that object has recently responded contingently to the infant's actions.
4. One of the best studies contesting this view found that mothers hyperarticulate vowels when talking to their babies, but not when talking to their pet cats or to fellow adults (Burnham, Kitamura, and Vollmer-Conna, 2002). However, consistent with the theory that motherese is just high pitched, emotional speech, the study also found more markers of emotion in the mothers' voices when they were addressing babies than when they were talking to cats.
5. Critics have questioned the evidence on which dual-process theory is based and identified inconsistencies among the various dual-process models on the market (Kruglanski and Gigerenzer, 2011). Evans and Stanovich (2013) offer a robust defense against these charges.
6. Some accounts suggest word learning involves specialized processes, but even these cognitive instinct accounts allow that associative learning plays a major role (Bloom, 2000; Kuhl, 2007).
7. Recent neuroanatomical studies challenge the established view that prefrontal cortex has expanded disproportionately in the course of primate evolution (Barton and Venditti, 2013; Gabi et al., 2016). For example, Gabi and colleagues found that, in humans, as in seven other primate species, the prefrontal cortex contains 8 percent of all cortical neurons. Consequently,

they propose that the complexity of human cognition is related not to the disproportionate size of our prefrontal cortex, but to the fact that it contains a larger absolute number of neurons than the prefrontal cortex of other primates. This view of brain evolution, although yet to become established, is in harmony with the view of cognitive evolution advanced in this book. They both suggest that, compared with our primate ancestors, humans have "more of the same."

4. CULTURAL LEARNING

1. "Long-term" is included in this definition of learning as an acknowledgement that there is no sharp line separating learning and perception, but the effects of the former tend to be more enduring.

5. SELECTIVE SOCIAL LEARNING

1. In other words, metacognitive social learning strategies are likely to promote what psychologists call "overimitation" or "overcopying"—the reproduction of irrelevant, or seemingly irrelevant, details of the model's behavior (see Chapter 6; Lyons, Young, and Keil, 2007; Whiten, McGuigan, Marshall-Pescini, and Hopper, 2009).
2. Evidence that nonhuman animals use explicit metacognition in solitary tasks (Beran, Smith, and Perdue, 2013; Beran et al., 2015), although controversial (Kepecs and Mainen, 2012; Shea et al., 2014), raises the possibility that some can also use metacognitive social learning strategies. However, the advantages of cook-like social learning strategies depend on their being constructed from the experience of many agents. Consequently, for nonhuman animals to reap the benefits of metacognitive social learning strategies, they would need not only to be capable of representing "who knows," but of communicating these beliefs to others. It is not inconceivable that this could be done without language, but it is unlikely that it could be done on any significant scale, and in a way that allows wisdom about "who knows" to accumulate over time.

6. IMITATION

1. Thanks to Luc-Alain Giraldeau, circa 1992, for the beautiful Lamborghini metaphor.

2. This is how the term "imitation" is now most commonly used in cognitive science. However, confusingly, it is sometimes used as a generic term referring to all kinds of social learning or behavioral copying.
3. Discussions of shared intentionality refer to "bird's eye," "third person," and "perspectival" representations of action (Tomasello, Carpenter, Call, Behne, and Moll, 2005; Tomasello and Moll, 2010). If these terms are seen as addressing the correspondence problem, discussions of shared intentionality offer a black box solution to that problem, similar to the one offered by the intermodal matching model. Like "supramodal representation," terms such as "perspectival representation" express confidence that something inside our heads is capable of converting the "seen but unfelt" into "the felt but unseen," but they don't tell us how the conversion is done.

7. MINDREADING

1. As a collectivist account, the cognitive gadget view has much in common with social constructivist accounts of mind reading (Carpendale and Lewis, 2004). However, unlike many social constructivists, and like theory-theory, the cognitive gadget view assumes that the processes involved in the development of mindreading are broadly rational and yield conceptual knowledge about the mind.
2. An intriguing follow-up study found that, in the test used by Southgate, Senju, and Csibra (2007), neurotypical adults show the same pattern of anticipatory looking as infants, but adults with Asperger's Syndrome look equally often to the left and to the right (Senju, Southgate, White, and Frith, 2009). According to the submentalizing interpretation, this suggests that, in adults, the anticipatory looking effect depends on explicit mentalizing or, more likely, that people with Asperger's Syndrome are less susceptible to distraction by social stimuli, such as an agent turning her back. The latter possibility is consistent with Senju and colleagues' observation that neurotypical adults spent more time than people with Asperger's Syndrome looking at the agent's face.
3. It is possible that everyday experience with arrows, in which interesting or important stimuli are more likely to be located near the head than the tail, results in habitual representation of what arrows can "see." This liberal version of the implicit mentalizing hypothesis is coherent, but it is not clear whether it is empirically testable. We know that *explicit* mentalizing can

be extended to virtually any object. If it is assumed that implicit mentalizing is also promiscuous—and given that we cannot, by definition, use verbal report to assess implicit mentalizing—there is a danger that implicit mentalizing hypotheses will become unfalsifiable. Under these circumstances, the dot perspective task would have no greater claim to demonstrate implicit mentalizing than, for example, the many experiments showing that eye and arrow stimuli induce involuntary shifts of attention (Guzzon, Brignani, Miniussi, and Marzi, 2010).

8. LANGUAGE

1. More attuned to the challenges of cultural learning, Simon Kirby and his colleagues advance a cultural account of the evolution of language which emphasizes selection pressure from the transmissibility of linguistic variants—how readily they can be learned by members of the incoming generation—rather than how readily they can be handled by the recipient's cognitive system once acquired (Tamariz and Kirby, 2016).
2. Both of these generalizations are supported by group-level analysis of functional imaging data. The latter is qualified by single-subject fMRI analyses focusing on what happens in individual brains. These suggest that Broca's area may contain two sets of regions, one involved more exclusively in language processing than the other (Fedorenko, Duncan, and Kanwisher, 2012).

9. CULTURAL EVOLUTIONARY PSYCHOLOGY

1. By definition, non-cultural social learning is not specialized for cultural inheritance (Chapter 4), but there is no reason to doubt that simple, ancient forms of social learning, such as stimulus enhancement and observational conditioning, can contribute to cultural inheritance (Heyes, 1993). They weren't made for the job, but they can still do it.
2. The regulative function of mindreading is an important target for future empirical research. My guess is that the acquisition of mindreading changes a range of processes in System 2—the part that most people think of as "the mind"—and has less impact on System 1 (see Chapter 3).
3. In principle, the problem could also be solved by selection of genetic mutations that speed up *all* sensorimotor learning, including that in which motor representations of body movement are linked with sensory representations of inanimate stimuli, and of nonmatching body movements.

This is a logical possibility, but: (1) it would be a case of imitation providing selection pressure for faster domain-general sensorimotor learning, rather than an example of the genetic assimilation of a domain-specific cognitive gadget, and (2) given the importance of sensorimotor learning in all motor control, across species, it is hard to believe that, if appropriate mutations were available, they would not have been selected long before imitation became important to humans. Furthermore, and crucially, given the need to ground genetic assimilation arguments in empirical evidence, I am not aware of any evidence that sensorimotor learning is faster in humans than in other animals.

Acknowledgments

I would like to thank not only those who influenced the grist of this book, the ideas it contains, but also the people who had such early and profound effects on the way I think that they shaped both the grist and the mills of this particular mind. Foremost among them is my brother, Vincent Heyes, who taught his little sister how to argue, and how to enjoy doing it—in the right company—above nearly all things. Somewhat later, the mill builders were my doctoral supervisor, Henry Plotkin, and postdoc advisors, Donald Campbell, Daniel Dennett, Anthony Dickinson, and Nicholas Mackintosh. In retrospect, I am glad there were such deep intellectual disagreements between them, but diving from the high board of evolutionary epistemology into the bracing water of associative learning did not always feel good. When I arrived in the Dickinson-Mackintosh lab in Cambridge, after two years in the United States with Campbell and Dennett, I felt sure I would convert—embrace empiricism, renouncing Mother Nature and all her works—or go insane. But the vision, depth of knowledge, and intellectual integrity of my advisors encouraged me to look for ways in which they could all be right. This book reports my progress so far in forging that synthesis.

Among those who shaped the grist of the book, I am especially grateful to the sharp minds with noble souls who read a whole draft of the manuscript and gave invaluable feedback: Martin Eimer, Chris Frith, Nick Shea, Kim Sterelny, and the members of a weekly discussion group, convened by Rachael Brown and Richard Moore, in the Philosophy Department at the Australian National University. (Sadly, I could not attend the discussion group, but I'm told that gales of vigorous dispute blew down the corridors whenever it was in session. I am trying, like Kim Sterelny, to take that as a Good Sign.) These generous first readers—alongside Ellen Clarke, Anthony Dickinson, Uta Frith, and John Pearce—have not only tutored and challenged me on the book's themes, but provided precious moral, practical, and intellectual support as I made the transition from "Experimental Psychologist" to "Theoretical Life Scientist."

On that road, between University College London and the University of Oxford, I gave up my lab, but I have not forgotten the demands of laboratory life. With ever more articles to write, grants to raise, and bean counters to satisfy—in addition to the usual academic workload of teaching and administration—it is very hard to find time for books. That is why I have worked long and hard to make this book short and, as far as possible, easy. It is written for accessibility to all those interested in human evolution—including anthropologists, archaeologists, biologists, economists, philosophers, and non-specialist readers—and to be short enough for swift consumption by people in psychology and neuroscience who rarely "do" books.

If I ever began to forget the rigors of laboratory life, I would be rudely reminded by Geoff Bird, Caroline Catmur, Richard Cook, Jane Leighton, Clare Press, and Elizabeth Ray. These friends and collaborators, with whom I developed the theory of imitation discussed in

Chapter 6, constantly remind me how difficult it is to design and implement a good experiment, and how rewarding it can be when done with people you admire.

The cognitive gadgets theory was at an early stage of gestation in 2011 when Robert Barton, Eva Jablonka, Alison Gopnik, Russell Gray, Arthur Robson, and Kim Sterelny came to Oxford for a term-long symposium on the evolution of human cognition. I am grateful to them for their stimulating company at that time, and to Henry VI of England, Archbishop Henry Chichele, and the Warden and Fellows of All Souls College, both for hosting the symposium, and providing an ideal academic environment in which to think about . . . almost anything.

This is my first, and quite possibly my only, book. I would like to thank Stefan Collini, Cecile Fabre, and Anthony Gottlieb for their guidance on the pleasures and perils of book writing, and the staff at Harvard University Press—including Janice Audet, Emeralde Jensen-Roberts, Ian Malcolm, and Stephanie Vyce—for their patient help with the nuts and bolts.

Among his many gifts, my partner, Martin Eimer, has remarkable place memory. Whenever our discussions have returned to a particular topic in this book, Martin has been able to recall where we last tackled the issue—for example, teleosemantics while walking in Highgate Woods, implicit mindreading outside a café in Genoa, linguistic universals on the way to a friend's wedding—and by that method to remember virtually everything we said. For this skill, but yet more for the acute insights and warm encouragement he has given in these unforgotten discussions, I am hugely grateful.

Index

M